Lecture Notes in Bioinformatics   **13254**

Subseries of Lecture Notes in Computer Science

More information about this subseries at https://link.springer.com/bookseries/5381

Mukul S. Bansal · Ion Măndoiu ·
Marmar Moussa · Murray Patterson ·
Sanguthevar Rajasekaran · Pavel Skums ·
Alexander Zelikovsky (Eds.)

Computational Advances in Bio and Medical Sciences

11th International Conference, ICCABS 2021
Virtual Event, December 16–18, 2021
Revised Selected Papers

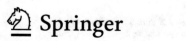

Springer

Editors
Mukul S. Bansal (iD)
University of Connecticut
Storrs, CT, USA

Marmar Moussa (iD)
University of Connecticut Health Center
Farmington, CT, USA

Sanguthevar Rajasekaran (iD)
University of Connecticut
Storrs, CT, USA

Alexander Zelikovsky (iD)
Georgia State University
Atlanta, GA, USA

Ion Măndoiu (iD)
University of Connecticut
Storrs, CT, USA

Murray Patterson (iD)
Georgia State University
Atlanta, GA, USA

Pavel Skums (iD)
Georgia State University
Atlanta, GA, USA

ISSN 0302-9743   ISSN 1611-3349 (electronic)
Lecture Notes in Bioinformatics
ISBN 978-3-031-17530-5   ISBN 978-3-031-17531-2 (eBook)
https://doi.org/10.1007/978-3-031-17531-2

LNCS Sublibrary: SL8 – Bioinformatics

This Springer imprint is published by the registered company Springer Nature Switzerland AG
The registered company address is: Gewerbestrasse 11, 6330 Cham, Switzerland

Preface

The 11th edition of the International Conference on Computational Advances in Bio and Medical Sciences (ICCABS 2021) was held in a virtual format during December 16-18, 2021. ICCABS has the goal of bringing together researchers, scientists, and students from academia, laboratories, and industry to discuss recent advances on computational techniques and applications in the areas of biology, medicine, and drug discovery.

There were 13 extended abstracts submitted in response to the ICCABS 2021 call for papers. Following a rigorous review process in which each submission was reviewed by two Program Committee members, the Program Committee decided to accept 10 extended abstracts for oral presentation and publication in the post-proceedings volume. The technical program of ICCABS 2021 included nine invited talks presented at the 11th Workshop on Computational Advances for Next Generation Sequencing (CANGS 2021), 16 invited talks presented at the 10th Workshop on Computational Advances in Molecular Epidemiology (CAME 2021), 12 invited talks presented at the 4th Workshop on Computational Advances for Single-Cell Omics Data Analysis (CASCODA 2021), and seven invited talks presented at the 1st Workshop on Advances in Systems Immunology (ASI 2021). Workshop speakers were invited to submit extended abstracts and, following the same review process used for the main conference, three additional extended abstracts were selected for publication in the post-proceedings volume. All extended abstracts included in the volume have been revised to address reviewers' comments.

The technical program of ICCABS 2021 also featured keynote talks by three distinguished speakers: Jan Korbel from the European Molecular Biology Laboratory, Heidelberg, gave a talk on "Multi-platform genomic sequencing reveals patterns of recurrent mutation associated with human diseases," Pavel Skums from Georgia State University gave a talk on "Phylodynamic analysis of tumors using single cell sequencing data," and Sharma Thankachan from the University of Central Florida gave a talk on "A Brief History of Genomic Data Indexing in Compressed Space." We would like to thank all keynote speakers and authors for presenting their work at the conference. We would also like to thank the Program Committee members and external reviewers for volunteering their time to review and discuss the submissions. Last but not least, we would like to extend special thanks to the Steering Committee members for their continued leadership, and to the webmaster, the finance chair, and the publicity chairs for their hard work in making ICCABS 2021 a successful event despite the ongoing COVID-19 pandemic.

July 2022

Mukul S. Bansal
Ion Măndoiu
Marmar Moussa
Murray Patterson
Sanguthevar Rajasekaran
Pavel Skums
Alexander Zelikovsky

Organization

Steering Committee

Srinivas Aluru	Georgia Institute of Technology, USA
Reda A. Ammar	University of Connecticut, USA
Tao Jiang	University of California, Riverside, USA
Vipin Kumar	University of Minnesota, USA
Ming Li	University of Waterloo, Canada
Sanguthevar Rajasekaran (Chair)	University of Connecticut, USA
John Reif	Duke University, USA
Sartaj Sahni	University of Florida, USA

General Chair

Sanguthevar Rajasekaran	University of Connecticut, USA

Program Chair

Sanguthevar Rajasekaran	University of Connecticut, USA

Workshop Chairs

Mukul S. Bansal	University of Connecticut, USA
Ion Măndoiu	University of Connecticut, USA
Marmar Moussa	University of Connecticut, USA
Murray Patterson	Georgia State University, USA
Pavel Skums	Georgia State University, USA
Alex Zelikovsky	Georgia State University, USA

Finance Chair

Reda A. Ammar	University of Connecticut, USA

Publicity Chairs

Orlando Echevarria	University of Connecticut, USA
Bob Weiner	University of Connecticut, USA

Webmaster

Zigeng Wang   University of Connecticut, USA

Program Committee

Sahar Al Seesi   Southern Connecticut State University, USA
Max Alekseyev   George Washington University, USA
Jaime Davila   Mayo Clinic, USA
Jorge Duitama   Universidad de los Andes, Colombia
Richard Edwards   University of New South Wales, Australia
Oliver Eulenstein   Iowa State University, USA
Danny Krizanc   Wesleyan University, USA
M. Oguzhan Kulekci   Istanbul Technical University, Turkey
Manuel Lafond   Université de Sherbrooke, Canada
Yuk Yee Leung   University of Pennsylvania, USA
Maria Poptsova   Moscow State University, Russia
Pavel Skums   Georgia State University, USA
Sing-Hoi Sze   Texas A&M University, USA
Sharma V. Thankachan   University of Central Florida, USA
Ugo Vaccaro   University of Salerno, Italy
Balaji Venkatachalam   Google, USA
Jianxin Wang   Central South University, China
Tianyu Wang   Amazon, USA
Fang Xiang Wu   University of Saskatchewan, Canada
Shibu Yooseph   University of Central Florida, USA
Shaojie Zhang   University of Central Florida, USA
Wei Zhang   University of Central Florida, USA
Cuncong Zhong   University of Kansas, USA

Additional Reviewers

Abedin, Paniz
Rondel, Filipp
Sahoo, Bikram
Shcheglova, Tatiana
Soliman, Ahmed
Vijendran, Sriram
Wang, Zigeng

Contents

Computational Advances in Bio
and Medical Sciences

Single Model Quality Estimation of Protein Structures via Non-negative Tensor Factorization

Kazi Lutful Kabir[1]([⊠])(ID), Manish Bhattarai[2], Boian S. Alexandrov[2], and Amarda Shehu[1](ID)

[1] Department of Computer Science, George Mason University,
Fairfax, VA 22030, USA
{kkabir,ashehu}@gmu.edu
[2] Theoretical Division, Los Alamos National Laboratory,
Los Alamos, NM 87545, USA
{ceodspspectrum,boian}@lanl.gov

Abstract. Finding the inherent organization in the structure space of a protein molecule is central in many computational studies of proteins. Grouping or clustering tertiary structures of a protein has been leveraged to build representations of the structure-energy landscape, highlight stable and semi-stable structural states, support models of structural dynamics, and connect them to biological function. Over the years, our laboratory has introduced methods to reveal structural states and build models of state-to-state protein dynamics. These methods have also been shown competitive for an orthogonal problem known as model selection, where model refers to a computed tertiary structure. Building on this work, in this paper we present a novel, tensor factorization-based method that doubles as a non-parametric clustering method. While the method has broad applicability, here we focus and demonstrate its efficacy on the estimation of model accuracy (EMA) problem. The method outperforms state-of-the-art methods, including single-model methods that leverage deep neural networks and domain-specific insight.

Keywords: Protein tertiary structure · Single model quality estimation · Tensor factorization

1 Introduction

The tertiary structure in which the amino acids constituting a protein molecule position themselves in three dimensions determines to a great extent the activities of a protein in the cell [9]. That is why the latest achievement of AlphaFold2 [15], which has already been employed to position about 58% of the total amino acids in human protein sequences [33], has been heralded as a great advance to obtain a mechanistic understanding of protein function. This achievement nonetheless ignores protein structure plasticity [24]. Methods that expand

© The Author(s), under exclusive license to Springer Nature Switzerland AG 2022
M. S. Bansal et al. (Eds.): ICCABS 2021, LNBI 13254, pp. 3–15, 2022.
https://doi.org/10.1007/978-3-031-17531-2_1

our view beyond one structure and account for the intrinsic ability of proteins to populate different functionally-relevant structures now abound [24]. Some methods sample the protein structure-energy landscape. Others use physics-based simulation to additionally reveal transitions between structures. Grouping/clustering computed structures of a protein has been used to build informative representations of the structure-energy landscape and expose the stable and semi-stable structural states that support the various activities of a protein in the cell. Work in [23] employs level sets to expose such states, whereas work in [16] leverages graph clustering and identifies macrostates by detecting communities in a graph embedding computed structures [18].

Over the years, we have proposed methods for grouping structures and supporting models of landscape-governed dynamics [16,19]. These methods have also been competitive in an orthogonal application setting, where the goal is to evaluate the quality of tertiary structures computed by one or more methods for a given protein and select a *best* structure [3,17]. This problem is also known as estimation of model accuracy (EMA) (model refers to a computed structure) or quality assessment (QA) and motivates much computational research [11].

In this paper we present a novel, tensor factorization-based method that organizes tertiary structures of a protein into groups. The method doubles as a non-parametric clustering method and so can broadly support various application settings. Here we focus and demonstrate its efficacy on EMA. As we expand further in Sect. 2, the method falls in the category of multi-model methods, as it extracts information from multiple structures/models. The method additionally computes an individual score for each structure that serves as a proxy of structure quality/accuracy. We show that the proposed method outperforms many state-of-the-art (SOTA) methods, including single-model methods. We first proceed with a review of related works.

2   Related Works

Early EMA methods were single-model. They arose before the term *model* in machine learning (ML) was popularized; model referred to a computed structure. Single-model methods provide one score per structure and first utilized molecular energy functions. These were not good proxies, though adding statistical terms offered some improvement [13]. Multi-model methods ignored scoring and clustered structures by similarity [38]. A combination of strategies were used to select a top cluster and then a best structure to offer as prediction. The landscape of multi-model methods is rich and now includes methods that consider structure energies to improve clustering [2]. Multi-model methods were superior for some time [27], until growth in structure databases facilitated single-model ML methods. Shallow ML methods focused on predicting a score for a given structure [26]. Then deep neural networks predicted increasingly accurate scores and are now SOTA for single-model methods in EMA [10].

We have investigated non-negative matrix factorization (NMF) to group structures for EMA. The NMF-MAD method in [3] decomposes a matrix of energy-based features of structures and outperforms multi-model methods. Work in [17] proposes SNMF-DS, which removes the reliance on features and factorizes a structure similarity matrix, utilizes symmetric NMF, and employs the eigen-gap statistic to automatically determine the number of groups. SNMF-DS outperforms NMF-MAD, MUFOLD-CL [38], and single-model methods, such as SBROD [20]. Matrix-factorization methods have been effective due to two reasons. First, they can handle sparse and highly imbalanced datasets, unlike most clustering methods. Second, the quality of computed tertiary structures has increasingly improved. In this paper we propose a factorization-based method but expand from matrix to tensor factorization. This proves more powerful, as we demonstrate its performance in Sect. 4. We now describe the method in greater detail.

3 Methodology

We refer to the proposed method as NTF-REL (non-negative tensor factorization with RESCAL) from now on. NTF-REL proceeds in four stages. Stage I organizes given structures into groups $\{G_i\}$ via tensor factorization. Stage II utilizes energies to rank the groups. Stage III partitions each group into subgroups and ranks them. Stage IV utilizes all this information to compute a score for each structure. A schematic is related in Fig. 1(a).

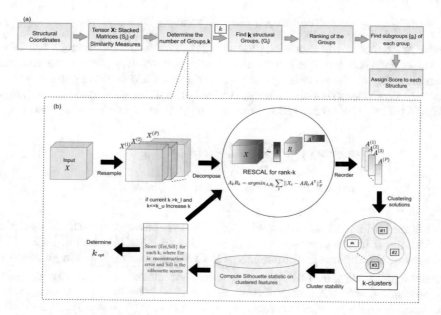

Fig. 1. (a) Schematic of the proposed method. (b) Finding the number of latent features with non-negative RESCAL factorization.

3.1   Stage I: From Structures to Groups

Different metrics for comparing two structures capture different aspects and often provide complementary information [29]. We form a tensor X by stacking (symmetric) similarity matrices $S_{n,n}^i$ obtained on n structures, where i refers to a particular metric. Entry (a, b) in S^i measures the similarity according to metric i between two structures at positions a, b in a list of n structures and $S_{a,a}^i = 1$. We use 5 popular metrics, RMSD, TM-score, GDT-TS, GDT-HA, and MaxSub score [29,31]. Since RMSD is a dissimilarity measure, we turn it into a similarity one as in $S_{a,b} = \frac{1}{\text{RMSD(a,b)}}$. Each S^i is a slice of the tensor. We use PyRosetta to obtain structures for a given protein, due to its popularity and ease of use; AlphaFold2 provides very few structures (about 45) in our experimentation with it. Each computed structure is stripped down to its main-chain carbon atoms (the CA atoms). This reduction brings down the cost of computing the tensor.

As Fig. 1(b) shows, the tensor X is then decomposed. We use the RESCAL tensor factorization approach [28] integrated with an automatic latent dimension determination method [32]. RESCAL was developed for extracting latent communities in relational data. Specifically, RESCAL factorizes tensors formed by a set of m stacked matrices of graphs (each graph has n nodes), $X^{n \times n \times m}$ into a factor matrix $A^{n \times k}$ and a core tensor $R^{k \times k \times m}$, where k is the latent dimension (or number of the latent communities/groups). The factorization solves the following optimization problem:

$$\text{argmin}_{A,R} \, \|X - R \times_1 A \times_2 A\|_F^2 \tag{1}$$

where $\times_i$ denotes the mode$-i$ product [21]. The extracted factors are interpretable; each column of A represents a latent community/group of objects, and each slice R_m of the core tensor R captures the relations among the groups at instance m. Considering the non-negativity of the data, we employ non-negative RESCAL [22]. The optimization with non-negativity constraints is given by,

$$\text{argmin}_{A,R_m} \quad \sum_m \left\|X_m - A R_m A^\top\right\|_F^2$$

$$\text{subject to} \quad \sum_j A_{ij} = 1, \text{ for } 1 \leq j \leq k; A, R \geq 0$$

Figure 1(b) shows our adaptation of RESCAL integrated with an algorithm to find the k latent groups, to which we refer as RESCAL-k [32]. RESCAL-k consists of the following components: (1) Custom Resampling: We generate an ensemble of X tensors, $[X^{(q)}]_{q=1,\ldots,P}$, where the means of these matrices equal to the original tensor X. Each of these tensors $X^{(q)}$ is built by perturbing each of the elements using random uniform noise, such that $X^{(q)} = X(\odot)\Delta_q$ (for details see [6]). (2) RESCAL Minimization: We use Frobenius norm-based multiplicative updates [28] to explore various numbers of latent features; k in an interval $[k_{min}, k_{max}]$, for each of the P generated random tensors $X^{(q)}$. The decomposed component A corresponds to the samples in reduced latent dimension $n \times k$

denoting the groups, whereas R is the $k \times k \times m$ relational tensor representing the group interactions. (3) Custom Clustering: For each $k \in [k_{min}, k_{max}]$, we cluster the set of the $n \times k$ latent components. To extract the latent dimension, we determine the dependency of the stability of the obtained clusters and the improvement of the reconstruction error on the latent dimension k. The final latent components, $\widetilde{A}$, are the medoids of the obtained stable clusters, with $\widetilde{R}$ denoting the corresponding mixing coefficients. The latent group estimation pipeline is based on the pyDRESCALk toolbox [7,8].

RESCAL-k employs Silhouette statistics [30] to determine the cluster stability for each k. The Silhouette parameter that quantifies the cluster stability is in the range $[-1, 1]$, where -1 corresponds to a bad clustering and 1 to perfect clustering. Figure 2 shows how using both Silhouette and reconstruction norm, we can determine the optimal k. The final representative A corresponding to the RESCAL factorization of X for k_{opt} estimated by RESCAL-k is then used to identify the best composition of the k groups identified, as illustrated in Fig. 2.

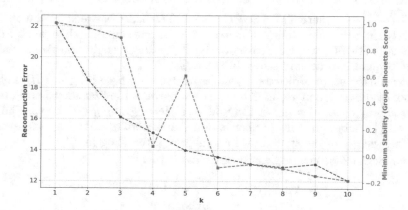

Fig. 2. We illustrate here the stability analysis for one of our protein targets, CASP target T1008-D1. Candidate values of k contain considerably larger gaps between the relative reconstruction error and the silhouette statistics with silhouette score $\geq$ 0.6. Out of the candidates (2, 3, and 5), we choose the one with the lowest reconstruction error (i.e., $k_{opt} = 5$ in this example).

3.2   Stage II: Ranking Groups

After determining the group composition using the matrix A identified as described above, we then rank the groups. Each group (of structures) is associated the average value over the Rosetta all-atom (score12) energies of the structures in the group. The groups are then ranked in ascending order of the group energy score; the group with the lowest score is the best-ranked group. The rank of a group G is denoted by R_G.

3.3   Stage III: Partitioning Groups into Subgroups

We hybridize the tensor-based approach above with graph clustering. We utilize work in [2] which embeds structure-energy pairs in a nearest-neighbor graph (using RMSD to identify nearest neighbors), over which it identifies local energy minima representing different energy basins and groups vertices into basins. In the interest of space, we spare the methodological details of how the minima and groupings are identified and refer the interested reader to work in [2]. Our adaptation here is not to apply this approach over all n structures, but instead to apply it to each group identified via tensor factorization in order to partition each group into "basins"; to which we refer more generally as subgroups.

3.4   Stage IV: Scoring Each Structure

To score each structure, we modify the strategy proposed in [1], which employs a model density score [37]. Let a structure x_i belong to a group G comprised of l structures. As in [1], we associate a density score ds_i with x_i as: $ds_i = \frac{\sum_{j=1}^{l} \text{TM} - \text{Score}_{i,j}}{l}$, where $\text{TM} - \text{Score}_{i,j}$ is the TM-Score between x_i and x_j ($1 \leq i, j \leq l$). In principle, different metrics can be measured; unlike work in [1], which uses RMSD, we use TM-score; its $[0, 1]$ range easily translates via the above formula into density scores in the range $[0, 1]$.

Our density score additionally utilizes information from Stages II-III. Let the number of subgroups in group G be z. The subgroups are first sorted and ranked in descending order of size. Let the rank of each subgroup $g \in G$ be r_g. The last d $\left(d = \lceil \frac{z}{3} \rceil\right)$ subgroups are further sorted (in ascending order) by the average potential energy of the structures in a subgroup, resulting in $r_g{}'$.

The modified structure density score $ds_i{}'$ is then:

$$ds_i{}' = \begin{cases} \frac{ds_i}{max(R_G) + r_g{}'} & \text{if } x_i \in \texttt{last } d \texttt{ subgroups} \\ \frac{ds_i}{R_G + r_g} & \texttt{otherwise} \end{cases}$$

Using these modified scores, we then assign weight/score w_i to each structure as in: $w_i = e^{ds_i{}'}$. Once the structures are weighted in this manner, the highest-weight structure is considered as the *best* structure.

3.5   Experimental Setup

We compare NTF-REL to representative SOTA methods: (1) Single-structure methods ProQ2 [36], ProQ3 [35], ProQ3D [34], and ProQ4 [25], and (2) recent NMF-based methods [3,4,17]. These NMF-based methods were shown to outperform MUFOLD-CL and other multi-model methods.

3.6   Dataset

As in [1,17], methods are evaluated on two datasets. The first, shown in Table 1 (left panel), contains 18 benchmark proteins of different folds and lengths (number of amino acids). The second dataset, shown in Table 1 (right panel), contains

10 targets selected from the free modeling category in CASP12 and CASP13; the list includes several hard targets. The Rosetta AbInitio protocol is used to generate 12,000 all-atom structures for each protein target. We note here that the goal is not to just generate one or a few structures, as now possible through AlphaFold2, but to obtain a broader view of the structure space containing alternative structures.

Tables 1 provide additional details for each dataset. Table 1 (left panel) shows the entry id of an experimentally known structure (ground truth) for each target in the Protein Data Bank (PDB) [5], the fold of the known structure, and the number of amino acids in the corresponding target. The minimum RMSD to the known structure in a dataset is shown in Column 6 and used to estimate the difficulty of a dataset for EMA. Targets where this value does not exceed 1Å are considered easy; those where this value does not exceed 3Å are considered medium; the rest are considered hard. Table 1 (right panel) lists similar information for the CASP targets. We note that in two cases, marked by asterisks, the known structure is only available on the CASP website.

Table 1. Left Panel: Benchmarks dataset (* denotes proteins with a predominant β fold and a short helix). The chain extracted from a multi-chain PDB entry is shown in parentheses. PDB ID, Fold, Length, and Min RMSD over a dataset to corresponding experimental structure are shown for each target. **Right Panel**: CASP dataset. CASP target IDs are shown in Column 2. PDB ID, Length, and Min RMSD over dataset to corresponding experimental structure are shown for each target. CASP targets with no experimentally-available structure in the PDB but only in the CASP website are marked by asterisks.

Difficulty	#	PDB ID	Fold	Length	Min RMSD (Å)
Easy	1	1ail	α	70	0.573
	2	1dtd(B)	$\alpha + \beta$	61	0.565
	3	1wap(A)	β	68	0.568
	4	1tig	$\alpha + \beta$	88	0.623
	5	1dtj(A)	$\alpha + \beta$	74	0.701
	6	1hz6(A)	$\alpha + \beta$	64	0.827
Medium	7	1c8c(A)	β^*	64	1.331
	8	2ci2	$\alpha + \beta$	65	1.581
	9	1bq9	β	53	1.308
	10	1hhp	β^*	99	1.761
	11	1fwp	$\alpha + \beta$	69	1.568
	12	1sap	β	66	2.031
	13	2h5n(D)	α	123	2.053
Hard	14	2ezk	α	93	3.475
	15	1aoy	α	78	3.496
	16	1aly	β	146	9.179
	17	1cc5	α	83	4.654
	18	1isu(A)	$coil$	62	5.912

#	Target ID	PDB ID	Length	Min RMSD (Å)
1	T1008-D1	6msp	77	1.542
2	T0886-D1	5fhy	69	5.102
3	T0953s1-D1	6f45	67	6.344
4	T0960-D2	6cl5	84	6.402
5	T0898-D2	**	55	6.598
6	T0892-D2	5nv4	110	6.950
7	T0953s2-D3	6f45	77	7.607
8	T0957s1-D1	6cp8	108	7.677
9	T0897-D1	**	138	9.638
10	T0859-D1	5jzr	113	10.268

3.7 Evaluation Metrics

Since NTF-REL assigns a score to each structure and so can also select a single structure as the best one, we evaluate its performance as an EMA method, as well as a single-structure selection method. First, we evaluate the quality of the scores assigned to structures by measuring the Pearson correlation between these scores and the true TM-Score from the *ground truth* (the experimentally-known structure for each target). We also measure loss as the difference in quality between the structure selected by a method and the best-quality structure in a dataset, with quality assessed by any of the three following metrics, RMSD, TM-Score, and GDT-TS, with respect to the experimentally-known structure.

4   Results

We present three sets of results, comparison with SOTA methods on target-wise correlation with respect to the true TM-Score, structure loss, and an analysis of statistical significance. We compare NTF-REL, SNMF-DS [17], NMF-MAD [3], ProQ2 [36], ProQ3 [35], ProQ3D [34], and ProQ4 [25] on the CASP and benchmark datasets.

4.1   Comparative Evaluation on Correlation with TM-Score

Table 2 relates the comparison on the benchmark and CASP targets. The top two predictions on each target are highlighted in boldface font. Table 2 shows that NTF-REL and ProQ4 outperform ProQ2, ProQ3, and ProQ3D on all benchmark and CASP targets. NTF-REL performs comparably to ProQ4, with differences often observed in the third digit after the decimal sign. In particular, both NTF-REL and ProQ4 achieve a Pearson correlation higher than 0.7 on 12/18 of the benchmark targets, respectively, and 8/10 and 10/10 of the CASP targets, respectively (with strictly no rounding). In many targets, both methods achieve or come very close to a Pearson correlation of 0.8.

Table 2. Target-wise Pearson correlation with respect to true TM-Score. Top two values are highlighted in boldface font.

Benchmark targets

Target-ID	NTF-REL	ProQ2	ProQ3	ProQ3D	ProQ4
1ail	**0.7821**	0.683	0.699	0.743	**0.787**
1dtj(A)	**0.802**	0.701	0.707	0.762	**0.807**
1dtd(B)	**0.7713**	0.675	0.683	0.733	**0.776**
1wap(A)	**0.7432**	0.658	0.665	0.713	**0.747**
1tig	**0.715**	0.624	0.634	0.689	**0.719**
1hz6(A)	**0.741**	0.647	0.657	0.714	**0.745**
1bq9	**0.688**	0.616	0.607	0.655	**0.697**
1c8c(A)	**0.728**	0.636	0.643	0.693	**0.733**
1fwp	**0.733**	0.641	0.646	0.702	**0.728**
1hhp	**0.679**	0.602	0.605	0.645	**0.683**
1sap	**0.7066**	0.617	0.621	0.673	**0.711**
2ci2	**0.746**	0.655	0.661	0.709	**0.741**
2h5n(D)	**0.7204**	0.623	0.637	0.685	**0.724**
1aoy	**0.686**	0.599	0.604	0.652	**0.677**
1aly	**0.652**	0.596	0.592	0.639	**0.661**
1cc5	**0.709**	0.627	0.625	0.684	**0.714**
1isu(A)	**0.6938**	0.607	0.611	0.679	**0.698**
2ezk	**0.667**	0.602	0.607	0.653	**0.675**

CASP targets

Target-ID	NTF-REL	ProQ2	ProQ3	ProQ3D	ProQ4
T0859-D1	**0.7031**	0.619	0.642	0.689	**0.717**
T0886-D1	**0.6972**	0.624	0.634	0.684	**0.704**
T0892-D2	**0.7044**	0.643	0.638	0.691	**0.719**
T0897-D1	**0.6896**	0.637	0.628	0.676	**0.703**
T0898-D2	**0.7203**	0.638	0.656	0.707	**0.734**
T0953s1-D1	**0.701**	0.632	0.603	0.652	**0.708**
T0953s2-D3	**0.718**	0.627	0.617	0.678	**0.725**
T0957s1-D1	**0.738**	0.602	0.635	0.696	**0.745**
T0960-D2	**0.7161**	0.624	0.646	0.667	**0.731**
T1008-D1	**0.7533**	0.643	0.648	0.701	**0.761**

4.2   Loss-Based Comparison

The above analysis suggests that ProQ3D and ProQ4 decidedly outperform ProQ2 and ProQ3, confirming findings reported in [12]. Therefore, we narrow further comparisons to ProQ3D and ProQ4. Since NTF-REL is a decomposion-based methods, like SNMF-DS and NMF-MAD, we add the latter two to the comparative evaluation on loss. As described in Sect. 3, we compute RMSD, TM-Score, and GDT-TS loss and relate these results in Table 3 on both benchmark and CASP targets.

Table 3. NTF-REL, SNMF-DS, ProQ3D, ProQ4, and NMF-MAD are compared on RMSD, TM-Score, and GDT-TS loss. Lowest loss per metric per target is highlighted in boldface font.

Benchmark targets

| ID | RMSD loss, TM-Score loss, GDT-TS loss | | | | |
	SNMF-DS	ProQ3D	ProQ4	NMF-MAD	NTF-REL
1ail	0.5084, 0.0655, 0.072	0.357, 0.042, 0.034	0.486, 0.063, 0.057	0.971, 0.1604, 0.1357	**0.1527, 0.03, 0.012**
1dtj(A)	0.1941, 0.0048, 0.0296	**0.125, 0.0118, 0.0057**	0.21, 0.0179, 0.0089	0.3345, 0.0782, 0.1081	0.166, **0.0043**, 0.026
1dtd(B)	0.3528, 0.0042, 0.0041	0.245, 0.0026, 0.0022	0.75, 0.0061, 0.0091	0.5915, 0.0329, 0.0451	**0.117, 0.0015, 0.0016**
1wap(A)	0.3425, 0.0288, 0.0166	0.352, 0.0277, 0.0245	0.423, 0.0311, 0.029	0.6219, 0.0531, 0.04	**0.2285, 0.021, 0.0123**
1tig	**0.0717, 0.003, 0.0053**	0.496, 0.0035, 0.0065	0.479, 0.0032, 0.0061	0.6569, 0.0469, 0.0483	0.72, 0.13, 0.091
1hz6(A)	**0.0936, 0.002, 0.0034**	0.397, 0.0031, 0.0042	0.291, 0.0025, 0.0039	0.809, 0.0415, 0.0352	0.1248, 0.005, 0.006
1bq9	1.1992, 0.1677, 0.1389	0.745, 0.112, 0.0896	**0.673, 0.103, 0.0749**	1.3089, 0.1167, 0.0755	1.6362, 0.226, 0.1875
1c8c(A)	0.7991, 0.1092, 0.086	0.521, 0.0953, 0.077	0.444, 0.0887, 0.069	1.092, **0.0596, 0.0429**	0.7991, 0.1092, 0.086
1fwp	0.5085, 0.0034, 0.0036	0.473, 0.0048, 0.0013	0.491, 0.0059, 0.0019	0.5319, 0.0471, 0.0616	**0.2658, 0.0019, 0.0023**
1hhp	2.1971, 0.0601, 0.0707	0.928, 0.073, 0.069	**0.77, 0.0467, 0.0503**	2.6835, 0.2939, 0.2828	2.3188, 0.0634, 0.075
1sap	0.5592, 0.074, 0.0417	0.719, 0.0637, 0.0398	0.875, 0.0714, 0.0416	2.075, 0.0989, 0.125	**0.3229, 0.045, 0.0248**
2ci2	0.3118, 0.007, 0.006	**0.213, 0.0056, 0.0042**	0.831, 0.013, 0.015	1.7897, 0.3246, 0.3462	0.3656, 0.01, 0.008
2h5n(D)	3.7028, 0.3178, 0.3215	0.917, 0.0475, 0.0276	**0.839, 0.0465, 0.0315**	3.3498, 0.0805, 0.0732	2.987, 0.267, 0.2708
1aoy	2.7896, 0.1136, 0.093	1.265, 0.0567, 0.0428	**1.074, 0.0511, 0.0431**	2.9788, 0.2918, 0.2788	2.346, 0.107, 0.089
1aly	5.7842, 0.0167, 0.0368	**2.674**, 0.0162, **0.027**	2.733, 0.0193, 0.0325	7.9939, 0.1411, 0.1635	2.912, **0.015**, 0.0345
1cc5	**0.4732, 0.048**, 0.0452	1.161, 0.0791, 0.0388	1.045, 0.0602, 0.0441	2.1843, 0.0565, 0.0573	0.539, 0.054, 0.0509
1isu(A)	2.9928, 0.2182, 0.2299	1.036, **0.072, 0.0705**	1.124, 0.0733, 0.0717	2.5552, 0.081, 0.0887	**0.8689**, 0.112, 0.135
2ezk	2.9154, 0.0188, 0.0177	0.729, 0.0027, 0.0063	**0.625, 0.0019, 0.0042**	3.5136, 0.0229, 0.0296	2.986, 0.021, 0.023

CASP targets

| Target ID | RMSD loss, TM-Score loss, GDT-TS loss | | | | |
	SNMF-DS	ProQ3D	ProQ4	NMF-MAD	NTF-REL
T1008-D1	0.3656, 0.007, **0.0011**	0.2838, 0.04, 0.035	0.326, 0.091, 0.088	1.0238, 0.0156, 0.0162	**0.2717, 0.006**, 0.005
T0886-D1	3.6714, **0.03**, 0.0362	0.983, 0.12, 0.112	**1.147**, 0.172, 0.153	2.5984, 0.0331, **0.029**	2.9813, 0.038, 0.034
T0953s1-D1	2.9398, **0.02, 0.0112**	**0.564**, 0.053, 0.041	1.179, 0.022, 0.019	2.613, 0.0225, 0.0223	3.3869, 0.0293, 0.0289
T0960-D2	1.8595, 0.0307, 0.0268	0.765, 0.13, 0.125	**0.634**, 0.067, 0.081	2.6181, **0.0182, 0.0178**	1.519, 0.031, 0.033
T0898-D2	1.4889, **0.003**, 0.0071	1.186, 0.019, 0.078	**0.917**, 0.0159, 0.068	2.3824, 0.0108, 0.0181	1.0799, 0.004, **0.0053**
T0892-D2	**0.9038, 0.0119**, 0.004	1.257, 0.189, 0.076	1.391, 0.263, 0.097	2.8416, 0.0242, 0.009	1.5471, 0.021, **0.0038**
T0953s2-D3	1.4223, **0.01**, 0.011	0.954, 0.161, 0.136	**0.818**, 0.0685, 0.078	1.8621, 0.0256, 0.0153	1.3667, 0.0218, 0.0108
T0897-D1	3.471, 0.0263, 0.018	0.973, 0.033, 0.013	0.849, 0.029, 0.011	2.9413, **0.0158, 0.009**	3.1845, 0.025, 0.0102
T0957s1-D1	1.18, 0.0027, 0.0047	1.161, 0.031, 0.096	1.294, 0.078, 0.173	1.6803, 0.018, 0.0076	**0.7426, 0.002, 0.0045**
T0859-D1	2.3755, 0.056, 0.045	**1.925**, 0.0752, 0.0853	1.972, 0.0734, 0.0771	3.5967, 0.0329, 0.0132	2.3317, **0.0265, 0.0124**

Table 3 shows the superiority of NTF-REL over other methods. For instance, on the benchmark targets, the RMSD loss incurred by NTF-REL is below $1\mathring{A}$ for 12/18 of the benchmark targets and under $2\mathring{A}$ for 6/10 of the CASP targets. The structure selected by NTF-REL has the minimum loss compared to the structure selected by other methods in terms of at least one of the three metrics (RMSD, TM-Score, and GDT-TS) on 8/18 of the benchmark targets and 6/10 of the CASP targets.

4.3   Statistical Significance Analysis

We carry out a statistical significance analysis on both TM-Score loss and GDT-TS loss combined over the benchmark and CASP datasets. We report the results of Friedman statistical tests with Hommel's post-hoc analysis [14]. We note that Friedman's test is ideal for conducting statistical significance of multiple methods contending over multiple test cases. The test is non-parametric and evaluates the null hypothesis (The null hypothesis states that there is negligible difference between the contending methods). Then, we conduct Hommel's post-hoc analysis to fully evaluate the performance of NTF-REL in comparison to other methods. The statistical tests are performed on all the 28 (benchmark and CASP) targets at $\alpha = 0.05$. The results are related in Table 4. The lowest average rank are reported in Columns 2 and 5 for TM-Score loss and GDT-Score loss, respectively. The best rank is achieved by NTF-REL on either TM-Score loss or GDT-TS loss. These results conclusively demonstrate the superiority of NTF-REL.

Table 4. Statistical significance of various methods over all 28 targets (benchmark and CASP) determined through Friedman's tests with Hommel's post-hoc analysis at $\alpha = 0.05$. The best rank on either TM-Score or GDT-TS loss is highlighted in boldface.

Method	TM-Score loss			GDT-TS loss		
	Avg. rank	p value	p Hommel	Avg. rank	p value	p hommel
NMF-MAD	3.107	0.063	0.0167	3.607	0.0425	0.0125
SNMF-DS	3.249	0.0562	0.0125	2.911	0.7037	0.025
ProQ3D	2.923	0.151	0.025	2.768	0.9663	0.05
ProQ4	2.893	0.177	0.05	2.965	0.6121	0.0167
NTF-REL	**2.322**	–	–	**2.75**	–	–

5   Conclusion

This paper presents a complete EMA framework that leverages a novel, tensor factorization-based method. The framework associates a score with an individual structure, so it has attributes of single-model EMA method. In addition, the method organizes structures into groups, so it has attributes of a multi-model method. The hybrid framework is shown to outperform various SOTA methods, including distance-based methods currently considered to be the most accurate.

The proposed tensor factorization method doubles as a non-parametric clustering method and so can support various structure-function studies requiring identification of structural macrostates. In this paper, for the purpose of a rigorous and targeted evaluation, we have restricted our attention to EMA and so many of our metrics of performance consider one, experimentally-available structure as the *ground truth*. However, as we make the case in Sect. 1, it is important to extend our attention beyond the single-structure view of proteins and evaluate, for instance, how the method proposed in this paper and others, can detect in the data multiple alternative, functionally-relevant structures. Our future work will investigate this setting, and it will additionally contribute curated benchmark datasets for rigorous evaluation.

In future work we will investigate additional settings, such as summarization of protein dynamics. The computation of the adjacency and degree matrices can be further expedited via proximity query data structures. Finding target-wise sub-spaces representative of a structure set may also prove informative.

Acknowledgment. This work is supported in part by NSF Grant No. 1900061 to AS. Resources were partly provided by the Los Alamos National Laboratory (LANL) Institutional Computing Program, which is supported by the DOE National Nuclear Security Administration under Contract No. DE-AC52-06NA25396 and LANL LDRD Grant No. 20190020DR. High-performance computations were run on Darwin, a LANL heterogeneous cluster for research computing, and on ARGO, a research computing cluster provided by the Office of Research Computing at George Mason University. This material is additionally based upon work by AS while serving at the National Science Foundation. Any opinion, findings, and conclusions or recommendations expressed in this material are those of the author and do not necessarily reflect the views of the National Science Foundation.

References

1. Akhter, N., Chennupati, G., Kabir, K.L., Djidjev, H., Shehu, A.: Unsupervised and supervised learning over the energy landscape for protein decoy selection. Biomolecules **9**(1), 607 (2019)
2. Akhter, N., Shehu, A.: From extraction of local structures of protein energy landscapes to improved decoy selection in template-free protein structure prediction. Molecules **23**(1), 216 (2018)
3. Akhter, N., Vangara, R., Chennupati, G., Alexandrov, B., Djidjev, H., Shehu, A.: Non-negative matrix factorization for selection of near-native protein tertiary structures. In: IEEE International Conference on Bioinformatics and Biomedicine (BIBM), pp. 70–73, San Diego, CA (2019)
4. Akhter, N., et al.: Improved protein decoy selection via non-negative matrix factorization. IEEE/ACM Trans. Comput. Biol. Bioinform. **19**(3), 1670–1682 (2021)
5. Berman, H.M., et al.: The protein data bank. Nucleic Acids Res. **28**(1), 235–242 (2000). https://www.rcsb.org/
6. Bhattarai, M., Chennupati, G., Skau, E., Vangara, R., Djidjev, H., Alexandrov, B.S.: Distributed non-negative tensor train decomposition. In: 2020 IEEE High Performance Extreme Computing Conference (HPEC), pp. 1–10. IEEE (2020)
7. Bhattarai, M., et al.: pyDRESCALk: python distributed non negative rescal decomposition with determination of latent features (2021)
8. Bhattarai, M., et al.: pyDNMFk: python distributed non negative matrix factorization (2021)
9. Boehr, D.D., Wright, P.E.: How do proteins interact? Science **320**(5882), 1429–1430 (2008)
10. Chen, X., Liu, J., Guo, Z., et al.: Protein model accuracy estimation empowered by deep learning and inter-residue distance prediction in CASP14. Sci. Rep. **11**, 10943 (2021)
11. Cheng, J., Choe, M., Elofsson, A.S., et al.: Estimation of model accuracy in CASP13. Proteins **87**(12), 1361–1377 (2021)
12. Cheng, J., Choe, M.H., Elofsson, A., et al.: Estimation of model accuracy in casp13. Proteins Struct. Funct. Bioinform. **87**(12), 1361–1377 (2019)

13. Felts, A.K., Gallicchio, E., Wallqvist, A., Levy, R.M.: Distinguishing native conformations of proteins from decoys with an effective free energy estimator based on the opls all-atom force field and the surface generalized born solvent model. Proteins Struct. Funct. Bioinform. **48**(2), 404–422 (2002)
14. Garcia, S., Herrera, F.: An extension on statistical comparisons of classifiers over multiple data sets for all pairwise comparisons. J. Mach. Learn. Res. **9**(12), 2677–2694 (2008)
15. Jumper, J., Evans, R., et al.: Highly accurate protein structure prediction with alphafold. Nature **596**(7873), 583–589 (2021)
16. Kabir, K.L., Akhter, N., Shehu, A.: From molecular energy landscapes to equilibrium dynamics via landscape analysis and markov state models. J. Bioinform. Comput. Biol. **17**(6), 1940014 (2019)
17. Kabir, K.L., Chennupati, G., Vangara, R., Djidjev, H., Alexandrov, B., Shehu, A.: Decoy selection in protein structure determination via symmetric non-negative matrix factorization. In: IEEE International Conference on Bioinformatics and Biomedicine (BIBM), pp. 23–28. Virtual (2020)
18. Kabir, K.L., Akhter, N., Shehu, A.: Unsupervised learning of conformational states present in molecular dynamics simulation data for summarization of equilibrium conformational dynamics. Biophys. J . **116**(3), 291a–292a (2019)
19. Kabir, K.L., Hassan, L., Rajabi, Z., Akhter, N., Shehu, A.: Graph-based community detection for decoy selection in template-free protein structure prediction. Molecules **24**(5), 854 (2019)
20. Karasikov, M., Pagès, G., Grudinin, S.: Smooth orientation-dependent scoring function for coarse-grained protein quality assessment. Bioinformatics **35**(16), 2801–2808 (2019)
21. Kolda, T.G., Bader, B.W.: Tensor decompositions and applications. SIAM Rev. **51**(3), 455–500 (2009)
22. Krompaß, D., Nickel, M., Jiang, X., Tresp, V.: Non-negative tensor factorization with rescal. In: Tensor Methods for Machine Learning, ECML Workshop, pp. 1–10 (2013)
23. Lei, J., Akhter, N., Qiao, W., Shehu, A.: Reconstruction and decomposition of high-dimensional landscapes via unsupervised learning. In: ACM SIGKDD International Conference on Knowledge Discovery & Data Mining, pp. 2505–2513, San Diego, CA (2020)
24. Maximova, T., Moffatt, R., Ma, B., Nussinov, R., Shehu, A.: Principles and overview of sampling methods for modeling macromolecular structure and dynamics. PLoS Comp. Biol. **12**(4), e1004619 (2016)
25. Menéndez Hurtado, D., Uziela, K., Elofsson, A.: A novel training procedure to train deep networks in the assessment of the quality of protein models (2019)
26. Mirzaei, S., Sidi, T., Keasar, C., Crivelli, S.: Purely structural protein scoring functions using support vector machine and ensemble learning. IEEE/ACM Trans. Compt. Biol. Bioinform. **16**(5), 1515–1523 (2016)
27. Moult, J., Fidelis, K., Kryshtafovych, A., Schwede, T., Tramontano, A.: Critical assessment of methods of protein structure prediction (casp)-round x. Proteins Struct. Funct. Bioinform. **82**, 1–6 (2014)
28. Nickel, M., Tresp, V., Kriegel, H.P.: A three-way model for collective learning on multi-relational data. In: Icml (2011)
29. Olechnovič, K., Monastyrskyy, B., Kryshtafovych, A., et al.: Comparative analysis of methods for evaluation of protein models against native structures. Bioinformatics **35**(6), 937–944 (2019)

30. Rousseeuw, P.J.: Silhouettes: a graphical aid to the interpretation and validation of cluster analysis. J. Comput. Appl. Math. **20**, 53–65 (1987)
31. Siew, N., Elofsson, A., Rychlewski, L., Fischer, D.: Maxsub: an automated measure for the assessment of protein structure prediction quality. Bioinformatics **16**(9), 776–785 (2000)
32. Truong, D.P., Skau, E., Valtchinov, V.I., Alexandrov, B.S.: Determination of latent dimensionality in international trade flow. Mach. Learn. Sci. Technol. **1**(4), 045017 (2020)
33. Tunyasuvunakool, K., Adler, J., Wu, Z., et al.: Highly accurate protein structure prediction for the human proteome. Nature **596**, 590–596 (2021)
34. Uziela, K., Menendez Hurtado, D., Shu, N., Wallner, B., Elofsson, A.: Proq3d: improved model quality assessments using deep learning. Bioinformatics **33**(10), 1578–1580 (2017)
35. Uziela, K., Shu, N., Wallner, B., Elofsson, A.: Proq 3: Improved model quality assessments using rosetta energy terms. Sci. Rep. **6**(1), 1–10 (2016)
36. Uziela, K., Wallner, B.: Proq2: estimation of model accuracy implemented in rosetta. Bioinformatics **32**(9), 1411–1413 (2016)
37. Wang, K., Fain, B., Levitt, M., Samudrala, R.: Improved protein structure selection using decoy-dependent discriminatory functions. BMC Struct. Biol. **4**(1), 1–18 (2004)
38. Zhang, J., Xu, D.: Fast algorithm for population-based protein structural model analysis. Proteomics **13**(2), 221–229 (2013)

Graph Representation Learning for Protein Conformation Sampling

Taseef Rahman[(⊠)], Yuanqi Du, and Amarda Shehu

Department of Computer Science, George Mason University, Fairfax, VA 22030, USA
{trahman2,ydu6,amarda}@gmu.edu

Abstract. Significant research on deep neural networks, culminating in AlphaFold2, convincingly shows that deep learning can predict the native structure of a given protein sequence with high accuracy. In contrast, work on deep learning frameworks that can account for the structural plasticity of protein molecules remains in its infancy. Many researchers are now investigating deep generative models to explore the structure space of a protein. Current models largely use 2D convolution, leveraging representations of protein structures as contact maps or distance matrices. The goal is exclusively to generate protein-like, sequence-agnostic tertiary structures, but no rigorous metrics are utilized to convincingly make this case. This paper makes several contributions. It builds on momentum in graph representation learning and formalizes a protein tertiary structure as a contact graph. It demonstrates that graph representation learning outperforms models based on image convolution. This work also equips graph-based deep latent variable models with the ability to learn from experimentally-available tertiary structures of proteins of varying lengths. The resulting models are shown to outperform state-of-the-art ones on rigorous metrics that quantify both local and distal patterns in physically-realistic protein structures. We hope this work will spur further research in deep generative models for obtaining a broader view of the structure space of a protein molecule.

Keywords: Graph representation learning · Protein structure plasticity · Conformation sampling

1 Introduction

Deep neural networks can learn directly from sequences and structures of proteins and accurately predict contacts of a novel amino-acid sequence. ResNet [21] was a precursor to AlphaFold2 [11], which recently showed that deep learning can predict tertiary structure from a sequence with high accuracy. This seminal development will support many structure-centric studies. Decades of research also show that a single-structure view ignores the inherent structural plasticity that proteins harness to regulate molecular interactions [2]. The Protein Data Bank (PDB) [1] is rich in proteins arrested in diverse biologically-active three-dimensional (3D)/tertiary structures [14]. Obtaining a broader view of the

M. S. Bansal et al. (Eds.): ICCABS 2021, LNBI 13254, pp. 16–28, 2022.
https://doi.org/10.1007/978-3-031-17531-2_2

structure space accessed by a protein is essential to advance our understanding of molecular mechanisms and support the development of therapeutics. Literature on methods for exploring the protein structure space is rich [15]. The majority operate under the umbrella of optimization and enhance the sampling capability of Monte Carlo- or Molecular Dynamics-based methods.

Momentum in deep generative models has recently spilled to protein structure modeling. A detailed review is beyond the scope of this paper, but we highlight here several lines of observation. Most related work confounds protein design, folding, and structure modeling. The objective is often not clearly stated. Largely, the goal is to show that generated tertiary structures *look like* structures of proteins, but rigorous evaluation is lacking. Most work does not connect with the long body of work and domain-specific insight that should inform and guide a quantitative evaluation of the realism of generated structures. Finally, most work leverages the generative adversarial network (GAN) architecture and builds over image-based convolution, representing a tertiary structure as a contact map or distance matrix, which are both 2D-based representations of a 3D structure. A more comprehensive survey of the landscape of deep generative models for protein structure modeling can be found in [10]. Several questions remain unanswered: (1) does 2D-based convolution suffice? (2) Which generative architecture is more powerful? (3) Are generated tertiary structures realistic?

In this paper, we provide answers to some of these questions and establish a clear, quantitative evaluation of the capability of deep generative models. First, we clarify our modest objective: Given known protein tertiary structures, learn to generate physically-realistic *conformations* of type-less amino-acid chains. Note how we utilize the concept of *a conformation*, which indicates a specific choice of representing a tertiary structure. A contact map is a conformation; a distance matrix is a conformation, resulting from a different representation choice. In this paper, our choice of conformation is a *contact graph*, which allows us to capture and leverage local and distal constraints inherent in a protein tertiary structure.

Guided by this objective, in this paper we carry out a detailed comparison that considers different architectures, such as GANs, Recurrent Neural Networks (RNNs), and Variational Autoencoders (VAEs). We evaluate the quality of generated conformations along several metrics that capture both local and distal patterns in (physically realistic) protein tertiary structures. Our most important contribution, from a methodological point of view, is our leveraging of graph representation learning, inspired by its promise for small molecule generation; *we present for the first time a graph-based model for protein conformation sampling.* We thus provide an additional dimension to the comparative evaluation, 2D convolution versus graph convolution models, and show through rigorous metrics that graph representation learning is more powerful to further advance protein conformation sampling. Before proceeding with methodological details, we first provide a concise review of the related work.

1.1   Related Work

Current deep generative methods represent a tertiary structure as a contact map or distance matrix, encoding the spatial proximity of pairs of amino acids (often collapsing an amino acid to its central carbon(CA) atom). Two CA's not further than 8Å in Euclidean space are considered in contact, and this is denoted by a 1 in the corresponding $[i, j]$ entry in the contact map/matrix, which is indexed by the position of CAs along the protein chain (from N- to C-terminus). A distance matrix records the actual Euclidean distances between CA pairs.

As the review in [10] shows, early works employed dihedral angle-based representations, but generated conformations contained many steric clashes, as such representations are under-constrained. State-of-the-art (SOTA) work uses contact maps or distance matrices and leverage 2D convolution (C)GANs popular in computer vision. Some methods specialize the loss function to focus the network to learn the symmetry of contact maps [9] or the sparse contacts/distances between amino acids far apart in the chain [5]. Recent work in [18] shows that, while GANs have become predominant frameworks, the quality of the contact or distance matrices they produce varies. Work in [18] proposes a Wasserstein GAN (WGAN) model that improves the quality of generated contact maps, and we use it as a baseline model here. We recall that WGAN uses 2D convolution, effectively treating a contact map or distance matrix as an image. Work in [18] also debuts metrics that quantify the local and distal patterns in physically-realistic tertiary structures, which we extend here to evaluate generated conformations of chains of varying lengths.

2   Methods

A conformation here is a contact graph, where amino acids become vertices, and contacts between pairs of CAs representing amino acids become edges. Vertices are labeled with the position of the corresponding CA atom in the chain. Using such graph representations allows us to adapt and evaluate GraphRNN and GraphVAE. As a baseline, to show the power of the graph representation, we utilize WGAN, which uses 2D convolution over contact maps. We summarize next GraphRNN and provide more details into GraphVAE, which we extend to learn from tertiary structures of varying-length protein chains.

GraphRNN: Consider an undirected graph $G = (A, E)$, where A represents the adjacency matrix of the graph, and E represents the edge attributes/features attached to each edge [22]. The goal of GraphRNN is to learn a distribution $p(G)$ over a set of observed graphs. GraphRNN defines a mapping f_S from graphs to sequences $f_S(G, \pi) = (S_1^\pi, ..., S_n^\pi)$, where π is a vertex ordering, and S_i^π denotes an adjacency vector representing edges between vertex v_i and previous vertices already in the graph. GraphRNN recasts $p(G)$ as the marginal distribution of the joint distribution $p(G, S^\pi)$: $p(G) = \sum_{S^\pi} p(S^\pi) \mathbb{1}[f_G(S^\pi) = G]$. Here, $p(S^\pi)$ is the distribution that needs to be learned. By approaching graph generation as a sequential process, GraphRNN formulates $p(S^\pi)$ as a product of conditional distributions over the elements; that is, $p(S^\pi) = \prod_{i=0}^{n} p(S_i^\pi | S_0^\pi, ..., S_{i-1}^\pi)$. Further details can be found in [22]. GraphRNN has been applied to learn grid networks, community (network) structures, and other real networks. It has also been employed in the generation of small molecules of a few dozen atoms. This paper evaluates it for the first time on the generation of protein structures.

GraphVAE: First proposed in [20], GraphVAE is able to generate probabilistic fully-connected graphs, where the output tensor consists of graphs of a fixed size. As illustrated in Fig. 1, GraphVAE, like traditional VAEs, consists of two separate networks, an encoder and a decoder that are trained simultaneously [6–8]. Let $G = (A, E, F)$ be a graph specified with its adjacency matrix A, edge features E, and node features F. The encoder maps a graph G to a latent space z by calculating the variational posterior $q(z|G)$. The decoder, a densely-connected layer, reconstructs the input graph in the form of $\tilde{G} = (\tilde{A}, \tilde{E}, \tilde{F})$ from the latent space z by a generative distribution $p(G|z)$. It subsequently employs a *reparameterization* trick that allows backpropagation of loss. Similar to VAEs [13], GraphVAE's loss also consists of Kullback-Liebler divergence (KL) loss and reconstruction loss. These respectively measure the divergence between $q(z|G)$ and $p(z)$ and the accuracy of the reconstructed graph $\tilde{G}$ as compared to the original graph G. One important aspect in which our GraphVAE differs from that of [20] is that we have circumvented the need for an explicit likelihood loss by making the encoder node-invariant through employing graph convolutions and node aggregation, inspired by the work in [4]. This also implies that the encoder can choose any one of the possible node orderings for a particular task, whereas the GraphVAE in [20] utilizes a max pooling matching algorithm [3].

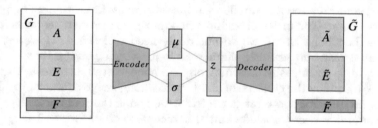

Fig. 1. Schematic shows the main components of GraphVAE.

vGraphVAE: We extend GraphVAE so that it can handle variable-size input contact graphs. We refer to the resulting model as *vGraphVAE*. We first find the longest chain in the input dataset. This becomes our target length (the target number of vertices in contact graphs). Each contact graph of a tertiary structure whose chain is shorter than the maximum length is padded with dummy vertices; the dummy vertices participate in no edges. During the interpretation of the output contact graph, we look for the first position where the aforementioned dummy vertex occurs, and we conduct the subsequent calculation up to that particular position.

Datasets: Our input dataset of tertiary structures is as in [16, 18]; 115, 850 tertiary structures are extracted from the PDB from entries listed in [16]. This set contains proteins of different lengths, which both GraphRNN and vGraph-VAE can handle, but WGAN cannot. We create three different views/training datasets of fixed-length chains to evaluate WGAN. We refer to these as FL = 16, FL = 64, and FL = 128 to indicate respective chain lengths of 16, 64, and 128 amino acids. As in [18], non-overlapping protein fragments of a given length l are extracted from chain 'A' for each protein structure starting at the first residue. The corresponding contact map is calculated and added to the training dataset for WGAN. This process is followed to obtain 115, 850 maps in the FL = 16 dataset and 98, 966 maps in the FL = 64 and FL = 128 datasets. Each model, WGAN, GraphRNN, and GraphVAE are separately trained on each of these datasets and then evaluated in each setting. Finally, GraphRNN and vGraph-VAE are compared to one another when the restriction of fixed-length is lifted. In that case, we consider the whole dataset of 115, 850 structures, from which we extract contact graphs. In each case, a $0.8 : 0.1 : 0.1$ split is followed for training, validation, and testing.

Metrics to Evaluate a Conformation: Inspired by the work in [18], we use domain-specific metrics to evaluate a generated conformation. First, we evaluate the presence of a backbone in a contact graph, which should be evident as an edge between two nodes corresponding to consecutive CAs. We sum up the number of such edges and refer to this metric as BackboneScore, thus summarizing each contact graph/conformation with a BackboneScore. Ideally, for a model trained over an $FL = l$ dataset, the average BackboneScore value over generated conformations should be $l - 1$. Smaller values indicate missing portions of the backbone. The order of amino acids in a given protein sequence tells us where the backbone is. So, we do not really need to learn it. However, it is a fundamental task and intrinsic characteristic in every structure that a powerful model ought to learn easily. Learning off-backbone, distal patterns is more challenging. We employ the concept of short- versus long-range contacts, inspired by the work in [18]. For short-range contacts, we count the number of (i, j) edges (which connect vertices labeled i and j) in a contact graph, where $1 < |j - i| \leq 4$ (the lower bound excludes the backbone). For long-range contacts, we restrict $|j - i| > 4$. Work in [18] considers contact maps of a fixed size. Here we evaluate models that can learn from a training dataset of varying-size contact graphs. So, we propose modifications of the above metrics to evaluate a contact graph by we normalizing them (dividing them by the number of vertices in a graph).

Metrics to Compare Distributions: The above metrics provide us with various ways to summarize a contact graph/conformation. They allow comparing the training to the generated dataset by comparing distributions of specific metrics. We make use of the Bhattacharya distance (BD) and the Earthmover Distance (EMD). BD [12] measures the distance between two distributions $p(x)$ and $q(x)$ defined over the same domain X. It is defined as $BD(p, q) = -\ln(BC(p, q))$. The Bhattacharaya coefficient $BC(p, q) = \sum_{x \in X} \sqrt{p(x)q(x)}$. BC varies from 0 to 1. BD varies from 0 to ∞. EMD [19] is also known as the Wasserstein distance. If the distributions are interpreted as two different ways of piling up a certain amount of dirt over the domain, EMD returns the minimum cost of turning one pile into the other. The cost is assumed to be the amount of dirt moved times the distance by which it is moved. EMD can be computed by solving an instance of the transportation problem, using any algorithm for minimum cost flow problem, such as the network simplex algorithm [19].

Implementation Details: All models are implemented using Pytorch [17]. Experiments are run on an NVIDIA Tesla V100 GPU, where an epoch of GraphRNN, vGraphVAE, and WGAN takes 19.51, 425.77, and 184.0 s, respectively.

3 Results

3.1 Experimental Setup

Part I of our experiments compare the performance of models trained on experimental protein tertiary structures of fixed-length chains. Trained WGAN, GraphRNN, and GraphVAE are compared on the quality of the dataset generated from each, using domain-agnostic and domain-specific metrics. Part II of our experiments compares the performance of models trained on experimental tertiary structures of varying-length chains. In each setting, we carry out three major analyses. We evaluate whether a trained model has learned to generate a backbone, short-range contacts, and long-range contacts. We make use of BD and EMD to compare distributions over the generated versus training dataset, as well as some visualizations of selected distributions. We arrest models at 10, 20, 30, and 50 epochs during the training process, so we can obtain a dynamic view. Inspection of loss over epochs shows that all models converge fast, within a few epochs (data not shown).

3.2 Evaluation of Models on Fixed-Length Chains

As described in Sect. 2, we design three experiments, constructing three separate training datasets; contacts graphs of 16, 64, and 128 vertices; for WGAN, these correspond to the number of rows and columns in contact maps. We refer to these datasets as FL = 16, FL = 64, and FL = 128, respectively. Each model is trained separately on each dataset and then utilized to generate contact graphs (of the corresponding size as in the respective training dataset), and the generated contact graphs are evaluated.

We first evaluate the presence of a backbone. Table 1 reports the average BackboneScore value over all instances generated by a model. Ideally, we expect average values to be nearly identical to FL − 1. Table 1 allows making several observations. The worst-performing model is GraphRNN. The average BackboneScore values it reports deviate significantly from the ideal ones. In comparison, WGAN and GraphVAE perform much better. However, WGAN's performance increases when the fragment length increases. The model has trouble on the shorter chains (see FL = 16). While better performing on the longer chains, FL = 64 and FL = 128, its performance varies significantly over training epochs;

Table 1. For each training dataset, the table reports the average of the generated distribution of BackboneScores from models arrested at 10, 20, 30, and 50 epochs.

FL	WGAN				GraphRNN				GraphVAE			
	10	20	30	50	10	20	30	50	10	20	30	50
128	126.98	125.26	114.23	71.37	6.17	6.09	6.17	6.29	126.86	126.90	126.91	126.89
64	63.0	61.45	62.04	55.87	5.39	5.56	5.14	5.43	62.93	62.95	62.96	62.93
16	1.82	0.84	0.31	0.04	2.60	2.83	2.81	3.81	14.99	14.99	14.99	14.99

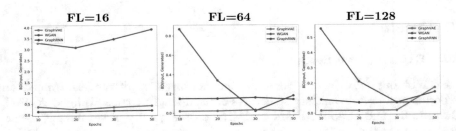

Fig. 2. The distribution of the number of short-range contacts in the generated dataset is compared to that in the training dataset via BD. The progression of BD's as a function of the number of training epochs for a specific model is tracked to show its impact on the quality of the generated dataset. This comparison is conducted separately for the models trained on the FL = 16, FL = 64, and FL = 128 datasets.

that is, this model is not stable with respect to learning the backbone. In contrast, GraphVAE is both the best performing, reporting values nearly identical to what is expected and stable over the training epochs.

The distribution of the number of short-range contacts over generated conformations is now compared to the corresponding distribution over the training data via BD and EMD. In the interest of space, we show here only the BD values, plotted in Fig. 2 for each model (on each training dataset) over epochs 10, 20, 30, and 50. Figure 2 shows that WGAN's generated distribution deviates significantly from the training distribution for the FL = 16 setting and does not improve with further training. The model improves with further training on the longer chains (FL = 64 and FL = 128). While GraphVAE and GraphRNN are close in performance, GraphVAE outperforms GraphRNN for the longer chains. There is an increase after 30 epochs on FL = 128, which suggests local instability. Altogether, the results suggest that GraphVAE is very effective at learning and reproducing the patterns of short-range contacts as in the training dataset. The EMD-based analysis confirms this (data not shown). We also show the actual distributions of short-range contacts over the training and generated dataset. The visualization in Fig. 3 is limited to models trained on the FL = 64 dataset and arrested at 50 training epochs. Figure 3 visually confirms the quantitative, rigorous analysis above, showing that the distribution of short-range contacts

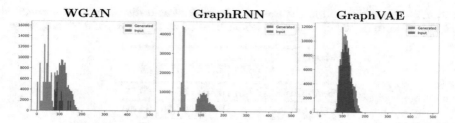

Fig. 3. The distribution of the number of short-range contacts corresponding to the generated dataset for each of the models is shown here. The visualization is limited to models trained on the FL = 64 dataset and arrested at 50 training epochs.

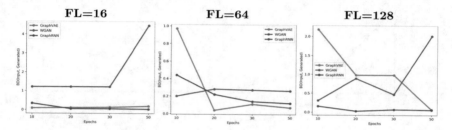

Fig. 4. The distribution of the number of long-range contacts in the generated dataset is compared to that in the training dataset via BD. The progression of BD's as a function of the number of training epochs for a specific model is tracked here to show its impact on the quality of the generated dataset. This comparison is conducted separately, for the models trained on the FL = 16, FL = 64, and FL = 128 datasets.

over the generated dataset is closer to the training distribution for GraphVAE, followed then by GraphRNN and by WGAN.

We repeat the above analysis but now for long-range contacts, again showing only BD values. Figure 4 shows that WGAN is not the best-performing model at any of the three training regimes. On FL = 16 and FL = 64, GraphVAE and GraphRNN performs similarly. On FL = 128, the performance of GraphVAE improves steadily over training epochs. The experiments suggest that GraphRNN is the more stable, followed closely by GraphVAE. The EMD-based analysis and the visualization of distributions confirm these observations (data not shown).

3.3   Evaluation of Models on Variable-Length Chains

The rest of the experiments compare GraphRNN to vGraphVAE. We fist evaluate the presence of a backbone. Table 2 reports the average of the normalized BackboneScore over all contact graphs in the generated dataset (x 100%). A dynamic view is provided over training epochs. Table 2 clearly shows that GraphRNN fails to produce even 50% of the backbone on average, whereas vGraphVAE produces over 95% of the backbone on average.

Table 2. The table reports the average of the generated distribution of normalized BackboneScores (x 100%) from each model arrested at 10, 20, 30, and 50 epochs.

Model	Epochs			
	10	20	30	50
GraphRNN	42.67	40.03	36.87	39.77
vGraphVAE	98.36	95.04	97.56	92.68

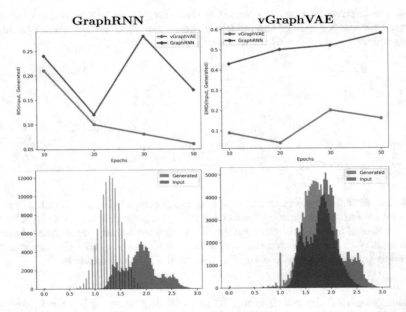

Fig. 5. Top panel: the distribution of normalized short-range contacts over the training and generated dataset are compared via (left) BD and (right) EMD. Bottom panel: the distributions are shown, superimposing the generated over the training distribution.

The distribution of normalized short-range contacts values over the training and generated dataset is compared for each model via BD and EMD and reported in the top panel of Fig. 5 over the training epochs. The bottom panel shows the actual distributions. Figure 5 clearly shows that vGraphVAE achieves the best performance (lowest BD and EMD values) over all training epochs, as well.

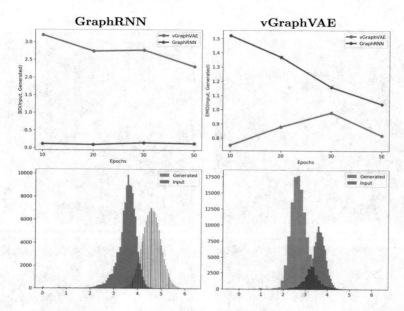

Fig. 6. Top panel: the distribution of normalized long-range contacts over the training and generated dataset are compared via (left) BD and (right) EMD. Bottom panel: the distributions are shown, superimposing the generated over the training distribution.

The distribution of normalized long-range values over the training and generated dataset is compared for each model via BD and EMD and reported in the top panel of Fig. 6 over the training epochs. The bottom panel shows the actual distributions. The top panel of Fig. 6 shows that both models achieve low BD and EMD values. vGraphVAE achieves lower EMD values. The bottom panel shows a better overlap between the input and generated distributions, which EMD seems to capture better.

Finally, we visualize some contact graphs selected at random over those generated by GraphRNN and vGraphVAE. Figure 7 draws them as contact maps, with bright yellow indicating edge/contact and dark blue indicating absence. The drawn contact maps are of high quality, with backbone, short-range, and long-range contacts, but those obtained by vGraphVAE are of higher quality. The visualization lends additional support to the conclusion that vGraphVAE is more effective than GraphRNN.

GraphRNN   vGraphVAE

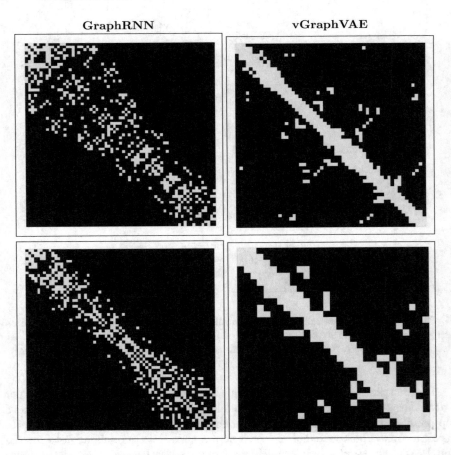

Fig. 7. Contact graphs are selected at random over generated data. Bright yellow indicates the presence of an edge/contact, and dark blue indicates the absence. (Color figure online)

4   Conclusion

This paper shows that the contact graph formalization is a very useful modality that advances representation learning for protein tertiary structures. A detailed comparison that pitches convolution-based to graph-based models, considers different architectures, such as GANs, RNNs, and VAEs, and evaluates the quality of protein tertiary structures along several informative metrics, suggests that the GraphVAE architecture is a good step towards generative models for protein structure generation. The ability to learn directly from experimental structures of proteins of varying lengths deposited in databases, such as the PDB, further advances the state of the art.

While the focus of this paper has been on improving the quality of generated structures, our end goal is to advance deep generative models, so that they can give us a complete and detailed view of the structure space of a given protein molecule. Further directions of work will include making such models sequence-specific. In this paper, we utilize the highly informative contact graph representation of a tertiary structure. Many algorithms exist to reconstruct tertiary structures from contact maps, and a natural direction is to expand our work to end-to-end models that readily provide tertiary structures.

Acknowledgment. This work is supported in part by NSF Grant No. 1907805, 1900061, and 1763233. Computations were run on ARGO, a research computing cluster provided by the Office of Research Computing at George Mason University, VA. This material is additionally based upon work by AS supported while serving at the National Science Foundation. Any opinion, findings, and conclusions or recommendations expressed in this material are those of the author and do not necessarily reflect the views of the National Science Foundation.

References

1. Berman, H.M., Henrick, K., Nakamura, H.: Announcing the worldwide protein data bank. Nat. Struct. Biol. **10**(12), 980–980 (2003)
2. Boehr, D.D., Nussinov, R., Wright, P.E.: The role of dynamic conformational ensembles in biomolecular recognition. Nature Chem. Biol. **5**(11), 789–96 (2009)
3. Cho, M., Sun, J., Duchenne, O., Ponce, J.: Finding matches in a haystack: a max-pooling strategy for graph matching in the presence of outliers. In: 2014 IEEE Conference on Computer Vision and Pattern Recognition, pp. 2091–2098 (2014). https://doi.org/10.1109/CVPR.2014.268
4. De Cao, N., Kipf, T.: Molgan: an implicit generative model for small molecular graphs. arXiv preprint arXiv:1805.11973 (2018)
5. Ding, W., Gong, H.: Predicting the real-valued inter-residue distances for proteins. Adv. Sci. **7**(19), 2001314 (2020)
6. Du, Y., Guo, X., Shehu, A., Zhao, L.: Interpretable molecule generation via disentanglement learning. In: Proceedings of the 11th ACM International Conference on Bioinformatics, Computational Biology and Health Informatics, pp. 1–8 (2020)
7. Du, Y., et al.: Graphgt: machine learning datasets for graph generation and transformation. In: Thirty-fifth Conference on Neural Information Processing Systems Datasets and Benchmarks Track (Round 2) (2021)
8. Guo, X., Du, Y., Zhao, L.: Property controllable variational autoencoder via invertible mutual dependence. In: International Conference on Learning Representations (2020)
9. Hang, H., Wang, M., Yu, Z., Zhao, X., Li, A.: GANcon: protein contact map prediction with deep generative adversarial network. IEEE Access **8**, 80899–80907 (2020)
10. Hoseini, P., Zhao, L., Shehu, A.: Generative deep learning for macromolecular structure and dynamics. Curr. Opinion Struct. Biol. **67**, 170–177 (2020)
11. Jumper, J., Evans, R., et al.: Highly accurate protein structure prediction with alphafold. Nature **596**(7873), 583–589 (2021). https://doi.org/10.1038/s41586-021-03819-2
12. Kailath, T.: The divergence and bhattacharyya distance measures in signal selection. IEEE Trans. Commun. Technol. **15**(1), 52–60 (1967)

13. Kingma, D.P., Welling, M.: Auto-encoding variational bayes. arXiv preprint arXiv:1312.6114 (2013)
14. Maximova, T., Carr, D., Plaku, E., Shehu, A.: Sample-based models of protein structural transitions. In: ACM Conference on Bioinformatics Computational Biology, and Health Informatics (BCB), pp. 128–137. Seattle, WA (2016)
15. Maximova, T., Moffatt, R., Ma, B., Nussinov, R., Shehu, A.: Principles and overview of sampling methods for modeling macromolecular structure and dynamics. PLoS Comp. Biol. **12**(4), e1004619 (2016)
16. Namrata, A., Raphael, E., Po-Ssu, H.: Fully differentiable full-atom protein backbone generation. In: International Conference on Learning Representations (ICLR) Workshops: DeepGenStruct (2019)
17. Paszke, A., et al.: Pytorch: an imperative style, high-performance deep learning library. In: Advances in Neural Information Processing Systems 32, pp. 8024–8035. Curran Associates, Inc. (2019). http://papers.neurips.cc/paper/9015-pytorch-an-imperative-style-high-performance-deep-learning-library.pdf
18. Rahman, T., Du, Y., Zhao, L., Shehu, A.: Generative adversarial learning of protein tertiary structures. Molecules **26**(5), 1209 (2021)
19. Rubner, Y., Tomasi, C., Guibas, L.J.: The earth mover's distance as a metric for image retrieval. Int. J. Comput. Vision **40**(2), 99–121 (2000). https://doi.org/10.1023/A:1026543900054
20. Simonovsky, M., Komodakis, N.: Graphvae: Towards generation of small graphs using variational autoencoders. In: International Conference on Artificial Neural Networks, pp. 412–422. Springer (2018). https://doi.org/10.1007/978-3-030-01418-6_41
21. Xu, J., McPartlon, M., Lin, J.: Improved protein structure prediction by deep learning irrespective of co-evolution information. Nature Mach. Intel. **3**, 601–609 (2020)
22. You, J., Ying, R., Ren, X., Hamilton, W., Leskovec, J.: GraphRNN: generating realistic graphs with deep auto-regressive models. In: International Conference on Machine Learning, pp. 5708–5717. PMLR (2018)

Excerno: Filtering Mutations Caused by the Clinical Archival Process in Sequencing Data

Audrey Mitchell[1], Marco Ruiz[1], Soua Yang[1], Chen Wang[2] (iD),
and Jaime Davila[1,2](✉) (iD)

[1] Department of Mathematics, Statistics and Computer Science, St Olaf College, Northfield,
MN 55057, USA
{mitch7,ruiz3,yang55,davila3}@stolaf.edu
[2] Department of Quantitative Health Sciences, Mayo Clinic, Rochester, MN 55905, USA
wang.chen@mayo.edu

Abstract. The accurate detection of mutations from clinical samples using Next Generation Sequencing (NGS) is of great importance in the clinical treatment of cancer patients. Clinical tests use archival pathology slides, which are preserved by Formalin-Fixation Paraffin Embedding (FFPE). The FFPE process introduces spurious $C > T$ mutations hindering accurate cancer diagnosis.

FFPE mutational artifacts occur in a well-defined pattern called a mutational signature. By quantifying the abundance of the FFPE mutational signature and using Bayes' formula we developed a method to filter FFPE artifacts. We implemented this method as the *excerno* package in the R statistical language.

We tested our method by generating mutations that follow the FFPE mutational signature and combining them with variants produced by other mutational signatures from the Catalog of Somatic Mutations in Cancer (COSMIC). First, we mixed an equal number of FFPE variants and mutations from a single COSMIC mutational signature and tested *excerno* across all of the 60 COSMIC mutational signatures. Our median sensitivity, specificity, and Area Under the Curve (AUC) were 0.89, 0.99, and 0.96 respectively. Furthermore, our performance characteristics decrease as a linear function of the similarity between the COSMIC and the FFPE mutational signatures ($R^2 = 0.90$). We also tested our method by mixing different proportions of mutations from the COSMIC and FFPE mutational signatures. As we increased the proportion of FFPE variants our sensitivity increased while our specificity decreased.

In conclusion, we developed and implemented *excerno*, an accurate method to filter FFPE artifactual mutations and characterized its performance characteristics using simulated datasets.

Keywords: Formalin-Fixation Paraffin-Embedded (FFPE) · Mutational signatures · Next Generation Sequencing (NGS)

A. Mitchell, M. Ruiz and S. Yang—Equal Contributions.

M. S. Bansal et al. (Eds.): ICCABS 2021, LNBI 13254, pp. 29–37, 2022.
https://doi.org/10.1007/978-3-031-17531-2_3

1   Introduction

The accurate detection of single nucleotide variants is of increasing importance in cancer since cancer sequencing tests are routinely in the clinical setting for diagnosis and therapy purposes. Clinical samples are generally preserved using Formalin-Fixation Paraffin-Embedding (FFPE), a tissue preserving process technique which allows accurate morphology assessment. Unfortunately, the FFPE process introduces an excess of C to T artificial mutations in Next Generation Sequencing (NGS) samples [1].

A mutational profile represents the distribution that a particular mutation arises in a particular dinucleotide context. A catalogue of 60 well-characterized mutational signatures corresponding to different mutational processes is available from the Catalogue Of Somatic Mutations in Cancer (COSMIC) [2, 3]. The COSMIC mutational signatures were initially established using Non-negative Matrix Factorization (NMF), an approach that approximates mutational profiles as a linear combination of a set of mutational signatures [4]. Multiple implementations of the NMF technique in the context of mutational signatures are available in the programming languages R and python [5, 6].

The mutational signature generated by the FFPE process from unrepaired NGS samples has been recently characterized in DNA samples [7]. This distinct mutational signature was also found in RNA sequencing (RNA-seq) from FFPE samples, and its abundance is correlated with the age of the tissue block [8]. This FFPE mutational signature has also been used to produce the original mutational profiles from DNA samples affected by FFPE artifacts [9]. Identifying and eliminating such artifactual mutations is of importance since they can hinder the correct diagnosis and treatment in the clinical practice.

In this work we leverage the FFPE mutational signature and Bayes' formula to build a method that establishes if a mutation originates from the FFPE process or not. We implemented this method as the *excerno* R package and tested it using simulated datasets where mutations following the FFPE mutational signature are added to mutations originating from a single COSMIC mutational signature.

2   *Excerno*: A Bayes Classifier Using Mutational Signatures

Our method can be summarized as follows. First, we compute the mutational profile from a sample, and use NMF or linear regression with positive coefficients to decompose such mutational profiles into COSMIC and FFPE mutational signature components. Afterwards we use Bayes formula to calculate the probability that a mutation is generated by a particular mutational signature (Fig. 1). Finally, we predict each mutation to be generated by the mutational signature that maximizes the posterior probability. In the remainder of this section we expand on the details for this method.

A Single Base Substitution (SBS) consists of a change of a single DNA base for a different nucleotide. Given the set of SBS from a sample we can construct the SBS mutational profile by counting the number of possible mutations ($C > A$, $C > G$, $C > T$, $T > A$, $T > C$, and $T > G$) in all possible 16 dinucleotide contexts. More formally a mutational profile is a vector of 96 components of the form:

$$\mathbf{prof}(x) := \left(n(x)_{A[C>A]A}, \ldots, n(x)_{T[C>A]T}, \ldots, n(x)_{A[T>G]A}, \ldots, n(x)_{T[T>G]T} \right).$$

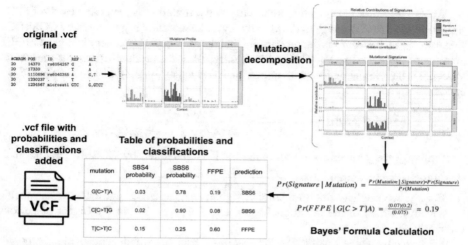

original .vcf file

.vcf file with probabilities and classifications added

Table of probabilities and classifications

mutation	SBS4 probability	SBS6 probability	FFPE	prediction
G[C>T]A	0.03	0.78	0.19	SBS6
C[C>T]G	0.02	0.90	0.08	SBS6
T[C>T]C	0.15	0.25	0.60	FFPE

$Pr(Signature \mid Mutation) = \frac{Pr(Mutation \mid Signature) \cdot Pr(Signature)}{Pr(Mutation)}$

$Pr(FFPE \mid G[C>T]A) = \frac{(0.07)(0.2)}{(0.075)} = 0.19$

Bayes' Formula Calculation

Fig. 1. Diagram of the *excerno* package workflow. Sample data is input as a vcf file, which is used to construct the mutational profile. Signature contributions are extracted using Non-Negative Linear Regression and posterior probabilities for each mutation and signature are calculated using Bayes' Formula. Mutations are classified according to the signature with the highest probability and a new vcf file with probabilities and classifications is returned.

Note that $n_{G[C>T]A}$ denotes the number of C > T changes that occur where G is the nucleotide before the mutation and T is the nucleotide after the mutation.

By dividing the mutational profile by the total number of mutations we obtain the mutational signature from sample x which we denote by:

$$\mathbf{sig}(x) := \frac{\mathbf{prof}(x)}{n} = (p(x)_{A[C>A]A}, \ldots, p(x)_{T[C>A]T}, \ldots, p(x)_{A[T>G]A}, \ldots, p(x)_{T[T>G]T}).$$

We will refer to a fixed mutation and its dinucleotide context by the letter V. The V-th component of *sig* is noted by sig_V and represents the frequency that the mutation appears in x. For example, if $V := G[C > T]A$ then $sig_{G[C>T]A}$ is the frequency that the mutation C > T appears preceded by G and followed by an A. Some examples of mutational signatures are the 60 COSMIC v.3.2 mutational signatures $c_1, \ldots, c_{60}$ and the FFPE mutational signature f [2, 7].

We are interested in calculating the posterior probability that V is generated by the FFPE mutational signature, which using Bayes formula becomes.

$$pr(FFPE \mid V) := pr(FFPE) \times \frac{pr(V \mid FFPE)}{pr(V)}.$$

To do that we would like to represent the mutational signature as a linear combination of the COSMIC and FFPE mutational signatures. More specifically we select k COSMIC mutational signatures, $\hat{c}_1, \ldots, \hat{c}_m$ and using linear regression with positive coefficients we obtain:

$$sig(x) \approx \beta_0(x)f + \beta_1(x)\hat{c}_1 + \ldots + \beta_k(x)\hat{c}_k$$

We can interpret $\beta_0(x)$ as the probability that a mutation is generated by the FFPE signature ($pr(FFPE)$) and β_i as the probability that a mutation is generated by COSMIC signature $\hat{c}_i$. Moreover f_V is the probability that the mutation V is generated by the FFPE mutational signature ($pr(V|FFPE)$). Finally, $sig_V(x)$ is the probability that the mutation V occurs in the sample x and can be approximated as $\beta_0(x)f_V + \beta_1(x)\widehat{c_1}_V + \cdots + \beta_k(x)\widehat{c_k}_V$ hence.

$$pr(FFPE|V) = \frac{\beta_0(x)f_V}{\beta_0(x)f_V + \beta_1(x)\widehat{c_1}_V + \ldots + \beta_k(x)\widehat{c_k}_V}.$$

Similarly the probability that a mutation V is generated by COSMIC mutational signature $\hat{c}_i$ can be computed by $\frac{\beta_i(x)\widehat{c_i}_V}{\beta_0(x)f_V + \beta_1(x)\widehat{c_1}_V + \ldots + \beta_k(x)\widehat{c_k}_V}$.

Finally, we predict which mutational signature generates a particular mutation V by selecting which of the mutational signatures $f, \hat{c}_1, \ldots, \hat{c}_k$ has the higher posterior probability.

We implemented our method as the *excerno* R package which is available at https://github.com/jdavilal/excerno. We used COSMIC mutational signatures version 3.2 and the recently discovered FFPE mutational signature [2, 7]. We leveraged the package *MutationalPatterns* R version 3.2.0 [5] and the function *fit_to_signatures* to obtain the mutational decomposition in COSMIC and FFPE mutational signatures.

3   Simulation and Evaluation Approach

To validate the *excerno* implementation and characterize its performance characteristics, we generated a simulated dataset where we mix mutations from a baseline COSMIC signature and the FFPE signature (Fig. 2).

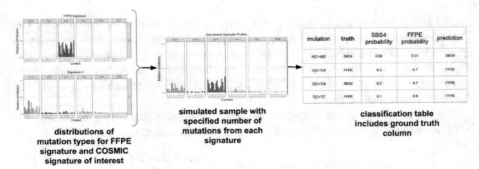

Fig. 2. Diagram of simulated data to calculate performance characteristics. The example depicts a simulated profile generated with 500 mutations from the FFPE signature and 500 mutations from SBS4.

Given a fixed mutational signature, we generated mutations and their dinucleotide context according to the mutational signature probabilities. We then mixed mutations generated from a single COSMIC mutational signature and mutations generated from the FFPE mutational signature. Each single dataset consists of 1000 mutations and for a single baseline COSMIC mutational signature we considered 10 independent replicates. For the first part of our paper we keep the number of FFPE and COSMIC mutations to be constant and equal to 500. For the latter part of our paper we consider different proportions of FFPE mutations from 5 to 95% in increments of 5%.

We ran *excerno* using the corresponding baseline COSMIC signature for the simulated dataset as a parameter. We plotted our Receiver Operating Characteristics (ROC) curves using the R package ROCR version 1.0–11. To measure the similarity between two mutational signatures we used the cosine similarity as implemented in the function *cos_sim* in the package *MutationalPatterns* version 3.2.0. We generated linear models using R version 4.1.0 and generated graphs using the R package tidyverse 1.3.1.

Our simulation and validation approaches were implemented as an R shiny application available at https://mitche7.shinyapps.io/excerno/. Such application allows the user to select a background COSMIC mutational signature as well as the number of mutations corresponding to the background and to the FFPE signature. The application then calculates the performance characteristics of the method, as well as the ROC curve.

4   Simulation Results

4.1   Performance Characteristics Across Different COSMIC Baseline Signatures

We considered two distinct datasets where we added an equal number of mutations from a baseline COSMIC mutational signature and the FFPE mutational signature. When using COSMIC mutational signatures SBS4 and SBS6 as a baseline, we obtained a sensitivity, specificity, and AUC of 0.90, 0.99, and 0.96 for SBS4 and 0.83, 0.73, and 0.82 for SBS6 (Fig. 3).

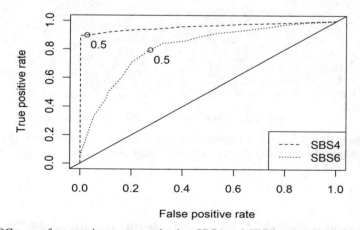

Fig. 3. ROC curve for mutations generated using SBS4 and SBS6 as baseline signatures. The AUC for SBS4 and SBS6 is 0.96 and 0.82.

The mutational profile of the filtered SBS4 dataset is very similar to SBS4. However, the mutational profile of the filtered SBS6 dataset differs from SBS6, especially in C >

T mutations preceded by an A or a T. We note that the probabilities associated with C > T mutations preceded by an A or a T are similar for the SBS6 and FFPE signatures.

We postulate that the performance characteristics of *excerno* depend on the similarity between the baseline and FFPE signatures, so we set-up simulated datasets where each of the 60 COSMIC mutational signatures were used as a baseline. Our median sensitivity, specificity, and AUC were 0.89, 0.99, and 0.96 with an IQR of 0.13, 0.09 and 0.07 respectively. By using the cosine similarity between the baseline and FFPE signatures we found a linear relationship between the similarity and the specificity, sensitivity and AUC (Fig. 4). The relationship between the AUC and the cosine similarity of both signatures can be written as $AUC = 1.00 - 0.29\ cos\_sim$, with an adjusted R^2 of 0.89. In particular the AUC of our method is lower than 0.8 for SBS11, SBS19, SBS23, SBS30, and SBS32. Finally, for every increase of 10% in cosine similarity between the baseline and FFPE signatures we obtained an average decrease in sensitivity and specificity of 3.7% and 3.9% respectively.

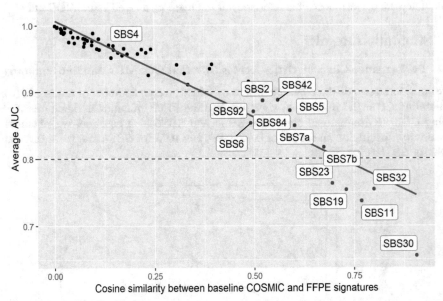

Fig. 4. Average AUC values plotted against the cosine similarity between baseline COSMIC and FFPE signatures. The following linear model was obtained $auc(x) = 1.00 - 0.29*cos\_sim(x, FFPE)$, $R^2 = 0.89$.

4.2 Performance Characteristics Across Different Percentages of FFPE

We added different FFPE mutation percentages to our baseline variants to establish a synthetic dilution experiment. We observed an increase of sensitivity and a decrease of specificity with higher percent of FFPE mutations (Fig. 5). When using SBS4 as our baseline we see a relatively constant sensitivity above 0.8 and specificity below 0.8

starting at 60% FFPE (Fig. 5). When using SBS6 as our baseline a sensitivity above 0.8 at lower than 55% FFPE and a specificity above 0.8 at higher than 60% FFPE (Fig. 5).

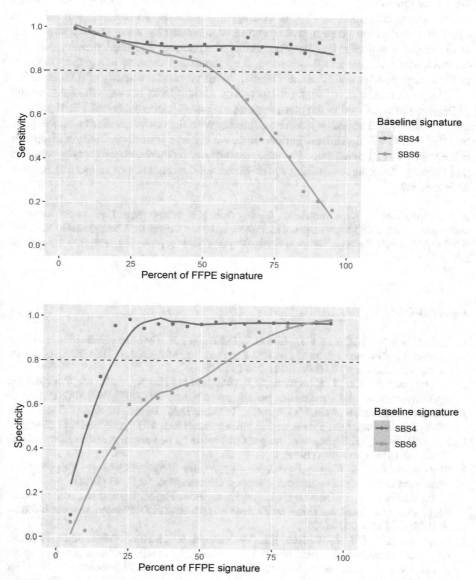

Fig. 5. Simulated samples of 1000 mutations with varying proportions of FFPE noise from 5% to 95% were mixed with mutation from baseline signature SBS4 and SBS6. Upper plot shows sensitivity against percentage of FFPE mutations with SBS4 in red and SBS6 in blue. Lower plot shows specificity against percentage of FFPE mutations with SBS4 in red and SBS6 in blue.

5 Conclusions

We present *excerno,* a method for filtering FFPE mutations that leverages the distinct FFPE mutational signature. Our method approximates the conditional probability of observing a mutation given a given mutational signature by using the decomposition of the mutational profile into known mutational signatures and implementing a Bayes classifier (10, 11). We tested our method on simulated datasets using the COSMIC mutational signatures and obtained a median sensitivity and specificity of 0.86 and 0.93.

The performance of our method depends on the similarity between the baseline and FFPE signatures, as well as the percentage of mutations generated by the FFPE process. Our specificity and sensitivity decrease linearly with the cosine similarity between the baseline and FFPE signatures. Furthermore, we observe lower specificity with higher percentage of FFPE mutations. For these reasons our method complements approaches that filter FFPE mutations making use of variant allele frequency or supporting number of reads (4, 7).

Acknowledgements. We acknowledge funding from Kay Winger Blair Endowment Award in Mathematics, the TRIO McNair Scholars Program and the Center for Undergraduate Research (CURI) from St Olaf College. We acknowledge Dr Asha Nair, for providing datasets that motivated this research.

References

1. Chen, G., Mosier, S., Gocke, C.D., Lin, M.T., Eshleman, J.R.: Cytosine deamination is a major cause of baseline noise in next-generation sequencing. Mol. Diagn. Ther. **18**(5), 587–593 (2014). https://doi.org/10.1007/s40291-014-0115-2
2. Alexandrov, L.B., Kim, J., Haradhvala, N.J., Huang, M.N., Tian Ng, A.W., Wu, Y., et al.: The repertoire of mutational signatures in human cancer. Nature **578**(7793), 94–101 (2020)
3. Tate, J.G., Bamford, S., Jubb, H.C., Sondka, Z., Beare, D.M., Bindal, N., et al.: COSMIC: the catalogue of somatic mutations in cancer. Nucleic Acids Res. **47**(D1), D941–D947 (2019)
4. Lee, D.D., Seung, H.S.: Learning the parts of objects by non-negative matrix factorization. Nature **401**(6755), 788–791 (1999)
5. Blokzijl, F., Janssen, R., van Boxtel, R., Cuppen, E.: MutationalPatterns: comprehensive genome-wide analysis of mutational processes. Genome Med. **10**(1), 33 (2018)
6. Islam, S.M.A., Wu, Y., Díaz-Gay, M., Bergstrom, E.N., He, Y., Barnes, M., et al.: Uncovering novel mutational signatures by *de novo* extraction with SigProfilerExtractor. bioRxiv. 2021:2020.12.13.422570 (2022)
7. Bhagwate, A.V., Liu, Y., Winham, S.J., McDonough, S.J., Stallings-Mann, M.L., Heinzen, E.P., et al.: Bioinformatics and DNA-extraction strategies to reliably detect genetic variants from FFPE breast tissue samples. BMC Genomics **20**(1), 689 (2019). https://doi.org/10.1186/s12864-019-6056-8
8. DiGuardo, M.A., Davila, J.I., Jackson, R.A., Nair, A.A., Fadra, N., Minn, K.T., et al.: RNA-Seq reveals differences in expressed tumor mutation burden in colorectal and endometrial cancers with and without defective DNA-mismatch repair. J. Mol. Diagn. **23**(5), 555–564 (2021)

9. Guo, Q., Lakatos, E., Al Bakir, I., Curtius, K., Graham, T.A., Mustonen, V.: The mutational signatures of formalin fixation on the human genome. bioRxiv. 2021:2021.03.11.434918 (2022)

Relabeling Metabolic Pathway Data with Groups to Improve Prediction Outcomes

Abdur Rahman M. A. Basher[1] and Steven J. Hallam[1,2]

[1] Graduate Program in Bioinformatics, University of British Columbia, Vancouver, BC V5Z 4S6, Canada
arbasher@student.ubc.ca, shallam@mail.ubc.ca
[2] Department of Microbiology & Immunology, University of British Columbia, Vancouver, BC V6T 1Z3, Canada

Abstract. Metabolic pathway inference from genomic sequence information is an integral scientific problem with wide ranging applications in the life sciences. As sequencing throughput increases, scalable and performative methods for pathway prediction at different levels of genome complexity and completion become compulsory. In this paper, we present reMap (relabeling metabolic pathway data with groups) a simple, and yet, generic framework, that performs relabeling examples to a different set of labels, characterized as groups. A pathway group is comprised of a subset of statistically correlated pathways that can be further distributed between multiple pathway groups. This has important implications for pathway prediction, where a learning algorithm can revisit a pathway multiple times across groups to improve sensitivity. The relabeling process in reMap is achieved through an alternating feedback process. In the first feed-forward phase, a minimal subset of pathway groups is picked to label each example. In the second feed-backward phase, reMap's internal parameters are updated to increase the accuracy of mapping examples to pathway groups. The resulting pathway group dataset is then be used to train a multi-label learning algorithm. reMap's effectiveness was evaluated on metabolic pathway prediction where resulting performance metrics equaled or exceeded other prediction methods on organismal genomes.

Keywords: Pathway group · Relabeling · Data augmentation · Correlated models · Metabolic pathway prediction · MetaCyc

This work was performed under the auspices of Genome Canada, Genome British Columbia, the Natural Sciences and Engineering Research Council (NSERC) of Canada, and Compute/Calcul Canada). ARMAB was supported by a UBC four-year doctoral fellowship (4YF) administered through the UBC Graduate Program in Bioinformatics.

M. S. Bansal et al. (Eds.): ICCABS 2021, LNBI 13254, pp. 38–50, 2022.
https://doi.org/10.1007/978-3-031-17531-2_4

1   Introduction

Biological systems operate on the basis of information flow between genomic DNA, RNA and proteins. Proteins catalyze most reactions resulting in metabolite production. Reaction sequences are called pathways when they contribute to a coherent set of interactions driving metabolic flux within or between cells. Inferring metabolic pathways from genomic sequence information is a fundamental problem in studying biological systems with important implications for our capacity to perceive, evaluate and engineer cells at the individual, population, and community levels of biological organization [4,8]. Over the past decade, the rise of next generation sequencing platforms has created a veritable tidal wave of organismal and multi-organismal genomes that must be assembled and annotated at scale without intensive manual curation. In response to this need, gene-centric and pathway-centric methods have been developed to reconstruct metabolic pathways from genomic sequence information at different levels of complexity and completion. The most common methods are gene-centric and involve mapping predicted protein coding sequences onto known pathways using a reference database (e.g. the Kyoto Encyclopedia of Genes and Genomes (KEGG) [6]. Alternative pathway-centric methods including PathoLogic [7] and MinPath [21] predict the presence of a given metabolic pathway based on heuristic or rule-based algorithms. While gene-centric methods are effective at producing parts list, they are unable to infer pathway presence or absence given a set of predicted protein coding sequences. Conversely, while pathway-centric methods infer pathway presence or absence given a set of predicted protein coding sequences, the development of reliable and flexible rule sets is both difficult and time consuming [20].

Machine learning methods aim to improve on heuristic or rule-based pathway inference through features engineering and algorithmic solutions to overcome noise and class imbalance. Basher and colleagues developed mlLGPR [12], a multi-label classification method that uses logistic regression and feature vectors inspired by the work of Dale and colleagues [3] to predict metabolic pathways from genomic sequence information at different levels of complexity and completion [12]. Recently, triUMPF [13,14] was proposed to reconstruct metabolic pathways from organismal and mutli-organismal genomes. This method uses meta-level interactions among pathways and enzymes within a network to improve the accuracy of pathway predictions in terms of communities represented by a cluster of nodes (pathways and enzymes). Despite triUMPF's predictive gains, its performance on pathway datasets left extensive room for improvement. Here, we present reMap that relabels each example with a new label set called "pathway group" or "group" forming a pathway group dataset which then can be employed by a suitable pathway prediction algorithm (e.g. leADS [19]) to improve prediction performance.

A subset of pathways in multiple organisms may be statistically correlated and this subset constitutes a group. Thus, the presence of a pathway entails the presence of a set of other correlated pathways. reMap performs an iterative procedure to group statistically related pathways into a set of "pathway groups"

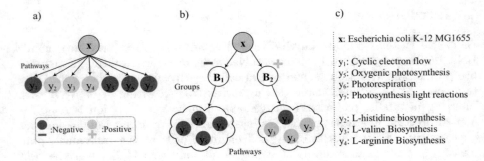

Fig. 1. Traditional vs proposed group-based pathway prediction methods. In the traditional method (a) pathways (i.e., y^{1-7}) are predicted for Escherichia coli K-12 MG1655, denoted by **x**, without considering any grouping of pathways. In contrast, the group-based pathway prediction method (b) uses a two step process. First, it predicts a set of positive groups (i.e., $\mathbf{B_2}$), then the pathways within these groups are predicted (depicted as a cloud glyph and true pathways are green colored). The description of symbols is provided in subfigure (c). (Color figure online)

using a correlation model (CTM, SOAP, and SPREAT see Appx. Section B [11]). reMap then annotates organismal genomes with relevant groups. Pathways in these groups are correlated and allowed to be inter-mixed across groups with different proportions, resulting in an overlapping subset of groups over a subset of pathways (i.e., non-disjoint). This has important implications for pathway prediction, where a learning algorithm can revisit a pathway multiple times across groups to improve sensitivity. Unlike mlLGPR [12] and triUMPF (Fig. 1a), group based pathway prediction requires two consecutive parts. First, a set of pathway groups are inferred. In the second, pathways in these groups are predicted (Fig. 1b).

reMap's pathway grouping performance was compared with other methods including MinPath, PathoLogic, and mlLGPR on a set of Tier 1 (T1) pathway genome databases (PGDBs), low complexity microbial communities including symbiont genomes encoding distributed metabolic pathways for amino acid biosynthesis [15], genomes used in the Critical Assessment of Metagenome Interpretation (CAMI) initiative [16], and whole-genome shotgun sequences from the Hawaii Ocean Time Series (HOTS) [17] following the genomic information hierarchy benchmarks initially developed for mlLGPR enabling more robust comparison between pathway prediction methods [12].

2   Method

In this section, we provide a general description of the reMap method, presented in Fig. 2. reMap is trained in two phases using an alternating feedback process: i)- feed-forward in Figs 2(b-d), consisting of three components: 1)- constructing pathway group, 2)- building group centroid, 3)- mapping examples to groups;

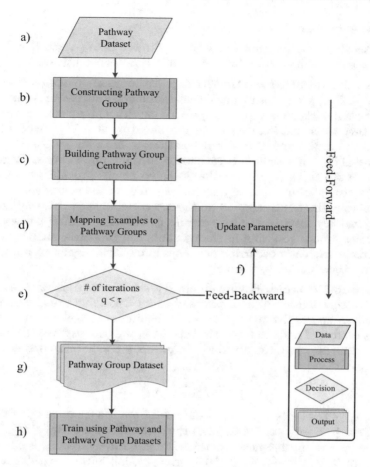

Fig. 2. A workflow diagram for reMAP. The relabeling process in reMap is achieved through an alternating feedback process. The feed-forward phase is composed of three components: (b) pathway group construction to build correlated pathway groups from pathway data (a), (c) building group centroid to estimate centroids of groups, and (d) mapping examples to groups. The feed-backward phase (f) optimizes reMap's parameters to increase accuracy of mapping examples to groups. The process is repeated τ ($\in \mathbb{Z}_{>1}$) times. If the current iteration q ($\in \mathbb{Z}_{>1}$) reaches the desired number of rounds τ, the training is terminated (e) and the pathway group dataset is produced (g) which can be used as inputs to a pathway inference algorithm (e.g. leADS [19]) to predict a set of pathways from a newly sequenced genome (h).

and ii)- feed-backward to update reMap's parameters in Fig. 2(f). After training is accomplished, a pathway group dataset is produced that can be used to predict metabolic pathways from a newly sequenced genome in Figs 2(g–h). Below, we discuss these two phases while the analytical expressions of reMap are explained in Appx. Section A [11].

2.1   Feed-Forward Phase

During this stage, each example in a given pathway data (Appx. Def. 1 [11]) is annotated with a subset of pathway groups in three consecutive steps:

Construct Pathway Group. In this step, pathways are partitioned into non-disjoint b ($\in \mathbb{Z}_{\geq 1}$) groups using any correlation models defined in Appx. Section B [11]. These models are equipped to provide us with a group correlation matrix and a pathway distribution over groups, denoted by $\Phi \in \mathbb{R}^{b \times t}$, where t corresponds to the total number of distinct pathways. Each entry $\Phi_{i,j}$ corresponds to the probability of assigning a pathway j to the group i. For each group in Φ, we retain the top k ($\in \mathbb{Z}_{\geq 1}$) pathways based on the probability scores. The trimmed Φ serves as an input to building centroids in the next step.

Modeling pathway distribution and group correlation in this way are motivated by two key intuitions. First, organisms encoding similar pathways may share similar groups resulting in shared statistical properties for those organisms. Second, frequently occurring pathways in multiple organisms imply a similar relative contribution to a group.

Build Group Centroid. Having obtained a set of groups, reMap determined the relative contribution of each pathway to its associated group's centroid in the Euclidean space. Estimating centroids requires representing pathways and groups as vectors of real numbers. For this, we apply pathway2vec [10] to obtain pathway features. Then, the centroid of a group, say s, is computed as:

$$\mathbf{c}_s = \frac{\alpha}{n_s} \sum_{j \in \mathbf{B}_{s,j} = +1} \frac{\mathbf{P}_j}{||\mathbf{P}_j||} \tag{2.1}$$

where $\mathbf{B}_s \in \{-1, +1\}^t$ is the group s obtained from the trimmed Φ_s after transforming it to $\{-1, +1\}^t$. $\mathbf{c}_s$ corresponds the centroid of the group s, $\mathbf{P} \in \mathbb{R}^{t \times m}$ is a pathway representation matrix obtained from pathway2vec, n_s is the number of pathways ($|\{\mathbf{B}_{s,j} = +1, \forall j \in t\}|$) in group s, $||.||$ is the length of a feature vector, and α ($\in \mathbb{R}_{>0}$) is a hyper-parameter determined by empirical analysis (16 in this work). The proposed Eq. 2.1 is based on the intuition that pathways associated with a group are semantically "close enough" to the center of the corresponding group, and the overlapping pathways among groups exhibit similar semantics with their associated groups. In addition to determining centroids, reMap also estimates a maximum number of expected groups to be annotated for a given example, indexed by i, using the cosine similarity metric [9]:

$$\widehat{\mathbf{D}}_i = \left\{ \mathbb{I}\left(\frac{\mathbf{c}_s^\top \tilde{\mathbf{c}}_s^{(i)}}{||\mathbf{c}_s|| \cdot ||\tilde{\mathbf{c}}_s^{(i)}||} \geq \upsilon \right) : 1 \leq s \leq b \right\}$$

$$\tilde{\mathbf{c}}_s^{(i)} = \frac{\alpha}{n_s} \sum_{j \in (\mathbf{Y}_{i,j} = +1 \wedge \mathbf{B}_{s,j} = +1)} \frac{\mathbf{P}_j}{||\mathbf{P}_j||} \tag{2.2}$$

where $\mathbb{I}(.)$ is an indicator function that results in either $+1$ or -1 depending on a user-defined threshold υ ($\in \mathbb{R}_{>0}$). $\mathbf{Y}_i \in \{-1, +1\}^t$ corresponds to pathways either present or absent for the ith example, indicated by $+1$ and -1, respectively. $\tilde{\mathbf{c}}_s^{(i)}$ represents the centroid of the group s calculated based on pathways that are associated with the group s and are present in ith example. $\tilde{n}_s$ is the number of pathways ($|\{\mathbf{Y}_{i,j} = +1 \wedge \mathbf{B}_{s,j} = +1, \forall j \in t\}|$) in group s. $\widehat{\mathbf{D}}_i \in \{+1, -1\}^b$ is a pre-optimized set of groups labelled for the ith example that will be used in the mapping step.

Map Pathways to Pathway Groups. This step maps an example to pathway groups, resulting in an optimized pathway group dataset $\widehat{\mathbf{D^{opt}}}$ ($\in \{+1, -1\}^{n \times b}$). Formally, let us denote a set of groups that are picked to label an example by $\mathcal{B}_{\mathbf{P}}^{(i)} \subseteq \arg\{\widehat{\mathbf{D}_{i,j}} = +1 : \forall j\}$ while the remaining unpicked groups is denoted by $\mathcal{B}_{\mathbf{U}}^{(i)}$, where $\widehat{\mathbf{D}}_i$ is obtained using Eq. 2.2. Both sets of groups are stored in $\mathcal{L}^{(i)} = \{\mathcal{B}_{\mathbf{P}}^{(i)} \cup \mathcal{B}_{\mathbf{U}}^{(i)}\}$. Then, reMap performs mapping in an iterative way, mirroring sequential learning and prediction strategy [18], where for each ith example, a group $\mathbf{B}_j$ at round q is either: i)-added to $\mathcal{L}^{(i)}$, indicated by $\mathcal{L}_q^{(i)} = \mathcal{L}_{q-1}^{(i)} \oplus \{\mathbf{B}_j : 1 < j \leq |\mathcal{B}_{\mathbf{U}}^{(i)}|\}$; or ii)- removed from the set of selected groups, represented by $\mathcal{L}_q^{(i)} = \mathcal{L}_{q-1}^{(i)} \ominus \{\mathbf{B}_j : 1 < j \leq |\mathcal{B}_{\mathbf{P}}^{(i)}|\}$. More specifically, at each iteration q, reMap estimates the probability of an example, given the selected groups that are obtained from the previous round $q - 1$, using the threshold closeness (TC) metric [2] as:

$$p\left(\mathbf{x}^{(i)} | \mathbf{H}_{q-1}^{(i)}, \mathcal{L}_{q-1}^{(i)}, \widehat{\mathbf{D}}_{i,j} = +1\right) = \frac{\bar{p}_{\mathbf{H}_{q-1}^{(i)}}\left(\widehat{\mathbf{D}_{i,j}} | \mathcal{L}_{q-1}^{(i)}, \mathbf{x}^{(i)}\right) G + \zeta}{Z} \qquad (2.3)$$

where $\mathbf{x}^{(i)} \in \mathbb{R}^r$ and r is the total number of enzymes, $G = 1 - \bar{p}_{\mathbf{H}_{q-1}^{(i)}}$ $(\widehat{\mathbf{D}_{i,j}} | \mathcal{L}_{q-1}^{(i)}, \mathbf{x}^{(i)})$ and $\widehat{\mathbf{D}_{i,j}} = +1$ if the group $\mathbf{B}_j$ is tagged with the ith example. $\mathbf{H}_{q-1}^{(i)}$ represents the history of prediction probability storing all $p(\widehat{\mathbf{D}_{i,j}} | \mathcal{L}_{q-1}^{(i)}, \mathbf{x}^{(i)})$ before the current iteration q while $\bar{p}_{\mathbf{H}_{q-1}^{(i)}}(\widehat{\mathbf{D}_{i,j}} | \mathcal{L}_{q-1}^{(i)}, \mathbf{x}^{(i)})$ is the average probability of classifying $\mathbf{x}^{(i)}$ to the group $\mathbf{B}_j$ over values in $\mathbf{H}_{q-1}^{(i)}$. The term ζ ($\in \mathbb{R}_{>0}$) is a smoothness constant and Z is a normalization constant. Note that TC is a class conditional probability density function that encourages correct class probability to be close to the true unknown decision boundary. Hence, this step will ensure the correct latent group to be assigned to the ith example. The parameter $p(\widehat{\mathbf{D}_{i,j}} | \mathcal{L}_{q-1}^{(i)}, \mathbf{x}^{(i)})$ can be estimated using Appx. Equation A.4 [11]. Afterwards, $\mathcal{L}^{(i)}$ will be updated either by adding or removing groups from a previous iteration. More details about this step is provided in Appx. Section A.1 [11].

2.2   Feed-Backward Phase

During this phase, reMap updates its internal parameters by enforcing four constraints: i)- similarity between groups and associated pathways; ii)- weights of pathways, in a group, should be close to each other; iii)- examples sharing similar pathways should share similar representations; and iv)- all reMap's parameters should not be too large or too small. These four constraints are important to allow smooth updates and mapping operations. More details are provided in Appx. Section A.2 [11].

2.3   Closing the Loop

The two phases are repeated for all examples in a given pathway data, until a predefined number of rounds τ $(\in \mathbb{Z}_{>1})$ is reached. At the end, a pathway group dataset is produced which consists of n examples with the assigned groups, i.e., $\widehat{\mathbf{D}^{opt}}$. After training is accomplished, a pathway group dataset is produced that can be used to predict metabolic pathways from a newly sequenced genome using an ML prediction method such as leADS [19].

3   Experiments

We evaluated reMap's performance on diverse pathway datasets traversing the genomic information hierarchy [12]: i)- T1 golden consisting of EcoCyc, Human-Cyc, AraCyc, YeastCyc, LeishCyc, and TrypanoCyc; ii)- BioCyc (v20.5 T2 & 3) [1]; iii)- *Symbionts* genomes of *Moranella* (GenBank NC-015735) and *Tremblaya* (GenBank NC-015736) encoding distributed metabolic pathways for 9 amino acid biosynthesis [15]; iv)- Critical Assessment of Metagenome Interpretation (CAMI) dataset composed of 40 genomes [16]; and v)- whole genome shotgun sequences from the Hawaii Ocean Time Series (HOTS) at 25 m, 75 m, 110 m (sunlit) and 500 m (dark) ocean depth intervals [17]. Information about these datasets is presented in Appx. Section C.1 [11].

Two experiments were conducted: i)- assessing the history probability and ii)- metabolic pathway prediction. The goal of the former test is to analyze the accumulated probability stored in $\mathbf{H}$ during the mapping process in the feed-forward phase for golden T1 datasets. We expect that few groups containing statistically related pathways will be annotated for T1 golden data. The metabolic pathway prediction test is followed to verify the quality of pathway groups for T1 golden, symbionts, CAMI, and HOTS data. For comparative analysis, reMap's performance on T1 golden datasets was compared to four pathway prediction methods: i)- MinPath version 1.2 [21], an integer programming based algorithm; ii)- PathoLogic version 21 [7], a symbolic approach that uses a set of manually curated rules to predict pathways; iii)- mlLGPR [12], a supervised multi-label classification and rich feature information algorithm, and iv)- triUMPF [13, 14], a non-negative matrix factorization and community detection based algorithm. Four metrics were used to report the performance of all pathway predictors for golden T1 and CAMI data: *average precision, average recall, average F1 score (F1)*, and *Hamming loss* as described in [12]. In addition, reMap's performance was compared to PathoLogic, mlLGPR, and triUMPF on mealybug

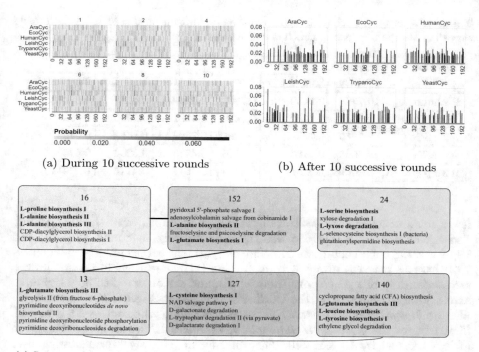

(a) During 10 successive rounds   (b) After 10 successive rounds

(c) Seven pathway groups that contain at least one amino acid biosynthesis pathways in these groups for Escherichia coli K-12 MG1655

Fig. 3. Figure 3a illustrates the history probability **H** during annotation of T1 golden data over 10 successive rounds while Fig. 3b shows the results after 10th round. Darker colors indicate higher probabilities of assigning groups to the corresponding data. Figure 3c shows six pathway groups and their correlations for Escherichia coli K-12 MG1655. Numbers at top boxes correspond to group indices. Edge thickness reflects the degree of associations between groups. Boldface text represent amino acid biosynthesis pathways.

symbionts and HOTS multi-organismal datasets. To construct pathway groups, we employed the correlated model SOAP using $b = 200$ groups.

reMap was written in Python v3 and is available under the GNU license at https://github.com/hallamlab/reMap. Unless otherwise specified all tests were conducted on a Linux server using 10 cores of Intel Xeon CPU E5-2650. For full experimental settings and additional tests, see Appx. Sections C and D [11].

3.1   Accumulated History Probability Analysis

Figure 3a shows **H** during the annotation process for the T1 golden data over 10 iterations. In the beginning, reMap attempts to select the maximum number of groups that may exist for each example. However, with progressive updates and calibration of parameters, reMap rectifies groups assignments where it picks

Table 1. Average F1 score of each comparing algorithm on 6 golden T1 data. Bold text suggests the best performance in each column.

Methods	Average F1 Score					
	EcoCyc	HumanCyc	AraCyc	YeastCyc	LeishCyc	TrypanoCyc
PathoLogic	0.7631	0.7460	0.7093	**0.7890**	0.6109	0.6447
MinPath	0.5161	0.4589	0.5489	0.4221	0.2990	0.3511
mlLGPR	0.7275	0.7468	**0.7343**	0.7392	**0.6220**	0.6768
triUMPF	0.8090	0.4703	0.4775	0.4735	0.5254	0.5266
reMap+SOAP	**0.8336**	**0.8285**	0.4764	0.4914	0.4144	**0.7305**

fewer relevant groups for each example. As an example, after the 10th round, Escherichia coli K-12 MG1655 was tagged with only 33 groups (Fig. 3b) and 18 of these groups contain amino acid biosynthesis pathways. Figure 3c shows 7 of these 18 pathway groups (Appx. Table 5 [11]). Pathways in these 7 groups are statistically related (Appx. Table 6 [11]), and are observed to be distributed across groups reflected by the thickness of edges in Fig. 3c. For example, *L-alanine biosynthesis II* pathway is present in groups indexed by 16 and 152. Similarly, for the pathway *L-glutamate biosynthesis III* which is represented in groups indexed by 13 and 140. This mixture of pathway representation over groups increases the chance of a pathway inference algorithm (e.g. leADS [19]) to revisit a true positive pathway multiple times across groups which may result in improved predictions as reported in the next section. This experiment shows that reMap is able to capture statistically relevant pathways and map related groups to each example with a high degree of correlation.

3.2 Metabolic Pathway Prediction

T1 Golden data. Table 1 shows that reMap+SOAP achieved competitive performance against the other methods in terms of average F1 score with optimal performance on EcoCyc (0.8336). However, it under-performed on AraCyc, YeastCyc, and LeishCyc, yielding average F1 scores of 0.4764, 0.4914, and 0.4144, respectively. Since reMap+SOAP was trained using BioCyc containing less than 1460 trainable pathways, pathways outside the training set will be neglected.

Symbionts data. The goal of this test is to evaluate reMap+SOAP performance on distributed metabolic pathways that emerge as a result of interactions between two or more organisms. We used the reduced genomes of *Moranella* and *Tremblaya* [15] as an established model for benchmarking. The two symbiont genomes in combination encode 9 intact amino acids biosynthesis pathways. All four pathway predictors were used to predict pathways on individual symbiont genomes and a composite genome consisting of both. While reMap+SOAP, triUMPF and PathoLogic predicted 6 of the expected amino acid biosynthesis pathways on the composite genome, mlLGPR was able to predict 8 pathways (Fig. 4).

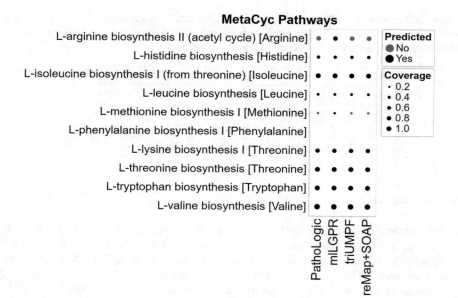

Fig. 4. Comparative study of predicted pathways for the composite genome between PathoLogic, mlLGPR, triUMPF, and reMap+SOAP. Black circles indicate predicted pathways by the associated models while grey circles indicate pathways that were not recovered by models. The size of circles corresponds the pathway coverage information.

We excluded phenylalanine biosynthesis (*L-phenylalanine biosynthesis I*) pathway from analysis because the associated genes were reported to be missing after initial gene prediction. Four predictors identified false positives for individual symbiont genomes in *Moranella* and *Tremblaya* although the pathway coverage information for both genomes was reduced in relation to the composite genome (Appx. Figure 9 [11]).

CAMI and HOTS data. For CAMI low complexity data [16], reMap+SOAP exceeded mlLGPR and triUMPF, achieving an average F1 score of 0.6125 in compare to 0.4866 for mlLGPR and 0.5864 for triUMPF (Table 2). For HOTS data [17], triUMPF, mlLGPR, and PathoLogic predicted a total of 58, 62, and 54 pathways, respectively, while reMap+SOAP inferred 67 pathways (see Appx. Section D.3 [11]) from a subset of 180 selected water column pathways [5]. None of the algorithms were able to predict pathways for *photosynthesis light reaction* and *pyruvate fermentation to (S)-acetoin* despite the abundance of these pathways in the water column. Absence of specific EC numbers associated with each pathway likely contributed to their absence using rule-based or ML prediction algorithms. Results from this experiment indicates that the proposed pathway group based approach, in particular reMap+SOAP increases pathway prediction performance relative to other methods used in isolation.

Table 2. Predictive performance of mlLGPR, triUMPF, and reMap+SOAP on CAMI low complexity data. For each performance metric, '↓' indicates the smaller score is better while '↑' indicates the higher score is better.

Metric	mlLGPR	triUMPF	reMap+SOAP
Hamming loss (↓)	0.0975	0.0436	**0.0407**
Average precision score (↑)	0.3570	0.7027	**0.7419**
Average recall score (↑)	**0.7827**	0.5101	0.5283
Average F1 score (↑)	0.4866	0.5864	**0.6125**

4 Conclusion

In this paper, we demonstrated that iteratively mapping examples to groups e.g. relabeling, using reMap increased pathway prediction performance. The reMAP method is based on the intuition that organisms sharing a similar set of metabolic pathways may exhibit similar higher-level structures or groups. The relabeling process in reMap is achieved through an alternating feedback process. In the first feed-forward phase, a minimal subset of pathway groups is picked to label each example. In the second feed-backward phase, reMap's internal parameters are updated to increase the accuracy of mapping examples to pathway groups. After training reMap, a pathway group dataset is produced that can be used to predict metabolic pathways for a newly sequenced genome.

We evaluated reMap's performance for the pathway prediction task using a corpus of experimental datasets and compared results to other prediction methods including PathoLogic, MinPath, mlLGPR, and triUMPF. Overall, reMap showed promising results in boosting prediction performance over ML-based algorithms, such as mlLGPR and triUMPF used in isolation. During benchmarking, we realized that reMap brings more frequent and sometimes irrelevant pathways, resulting in a significant performance loss on some T1 golden data, such as AraCyc. A possible treatment would be adding constraints in the form of associations among enzymes and pathways as applied in triUMPF. However, this may lead to sensitivity loss. Another approach is to combine both graph-based and group-based strategies to predict pathways. Future development efforts will explore this dual approach to improve pathway prediction performance with emphasis on multi-organismal genomes encoding distributed metabolic processes.

References

1. Caspi, R., Billington, R., Foerster, H., et al.: Biocyc: online resource for genome and metabolic pathway analysis. FASEB J. **30**(1 Supplement), lb192-lb192 (2016)
2. Chang, H.S., Learned-Miller, E., McCallum, A.: Active bias: training more accurate neural networks by emphasizing high variance samples. In: Advances in Neural Information Processing Systems, pp. 1002–1012 (2017)

3. Dale, J.M., Popescu, L., Karp, P.D.: Machine learning methods for metabolic pathway prediction. BMC Bioinform. **11**(1), 1 (2010)
4. Hahn, A.S., Konwar, K.M., Louca, S., et al.: The information science of microbial ecology. Curr. Opin. Microbiol. **31**, 209–216 (2016)
5. Hanson, N.W., Konwar, K.M., Hawley, A.K., et al.: Metabolic pathways for the whole community. BMC Genomics **15**(1), 1 (2014). https://doi.org/10.1186/1471-2164-15-619
6. Kanehisa, M., Furumichi, M., Tanabe, M., et al.: KEGG: new perspectives on genomes, pathways, diseases and drugs. Nucleic Acids Res. **45**(D1), D353–D361 (2017)
7. Karp, P.D., Latendresse, M., Paley, S.M., et al.: Pathway tools version 19.0 update: software for pathway/genome informatics and systems biology. Briefings Bioinform. **17**(5), 877–890 (2016)
8. Lawson, C.E., Harcombe, W.R., Hatzenpichler, R., et al.: Common principles and best practices for engineering microbiomes. Nat. Rev. Microbiol. **17**(12), 725–741 (2019)
9. Luo, C., Zhan, J., Xue, X., Wang, L., Ren, R., Yang, Q.: Cosine normalization: using cosine similarity instead of dot product in neural networks. In: Kůrková, V., Manolopoulos, Y., Hammer, B., Iliadis, L., Maglogiannis, I. (eds.) ICANN 2018. LNCS, vol. 11139, pp. 382–391. Springer, Cham (2018). https://doi.org/10.1007/978-3-030-01418-6_38
10. MA Basher, A.R., Hallam, S.J.: Leveraging heterogeneous network embedding for metabolic pathway prediction. Bioinformatics **37**(6), 822–829 (2020). https://doi.org/10.1093/bioinformatics/btaa906
11. Basher, A.R.M., Hallam, S.J.: Relabeling metabolic pathway data with groups to improve prediction outcomes. bioRxiv (2021). https://doi.org/10.1101/2020.08.21.260109
12. Basher, A.R.M., McLaughlin, R.J., Hallam, S.J.: Metabolic pathway inference using multi-label classification with rich pathway features. PLoS Comput. Biol. **16**(10), 1–22 (2020)
13. Basher, A.R.M., McLaughlin, R.J., Hallam, S.J.: Metabolic pathway prediction using non-negative matrix factorization with improved precision. J. Comput. Biol. **28**(11), 1075–1103 (2021)
14. Mohd Abul Basher, A.R., McLaughlin, R.J., Hallam, S.J.: Metabolic pathway prediction using non-negative matrix factorization with improved precision. In: Jha, S.K., Măndoiu, I., Rajasekaran, S., Skums, P., Zelikovsky, A. (eds.) ICCABS 2020. LNCS, vol. 12686, pp. 33–44. Springer, Cham (2021). https://doi.org/10.1007/978-3-030-79290-9_4
15. McCutcheon, J.P., Von Dohlen, C.D.: An interdependent metabolic patchwork in the nested symbiosis of mealybugs. Curr. Biol. **21**(16), 1366–1372 (2011)
16. Sczyrba, A., Hofmann, P., Belmann, P., et al.: Critical assessment of metagenome interpretation-a benchmark of metagenomics software. Nat. Methods **14**(11), 1063 (2017)
17. Stewart, F.J., Sharma, A.K., Bryant, J.A., et al.: Community transcriptomics reveals universal patterns of protein sequence conservation in natural microbial communities. Genome Biol. **12**(3), R26 (2011)
18. Sutskever, I., Vinyals, O., Le, Q.V.: Sequence to sequence learning with neural networks. In: Advances in Neural Information Processing Systems, pp. 3104–3112 (2014)
19. Hallam Lab: leADS. https://github.com/hallamlab/leADS

20. Toubiana, D., Puzis, R., Wen, L., et al.: Combined network analysis and machine learning allows the prediction of metabolic pathways from tomato metabolomics data. Commun. Biol. **2**(1), 214 (2019)
21. Ye, Y., Doak, T.G.: A parsimony approach to biological pathway reconstruction/inference for genomes and metagenomes. PLoS Comput. Biol. **5**(8), e1000465 (2009)

MELEPS: Multiple Expert Linear Epitope Prediction System

An-Chi Shau[1], Nai-Shuan Hwang[1], Shu-Yu Chang[1], Hsin-Yiu Chou[2], and Tun-Wen Pai[1,3(✉)]

[1] Department of Computer Science and Engineering, National Taiwan Ocean University, Keelung, Taiwan
`twp@ntut.edu.tw`
[2] Department of Aquaculture, National Taiwan Ocean University, Keelung, Taiwan
[3] Department of Computer Science and Information Engineering, National Taipei University of Technology, Taipei, Taiwan

Abstract. Taiwan is one of the most important fishery countries in the world due to its leading fishery breeding and farming technologies. However, high-density fishery farming environments are vulnerable to bacteria or viruses and would cause serious losses. Predicting epitope binding segments from pathogenic bacteria is the first step for vaccine and drug development, and bioinformatics technologies could provide effective approaches to facilitate effective prediction of epitope segments. This study integrated six linear epitope prediction systems for prediction of highly antigenic segments through a weighted voting mechanism. Testing datasets were retrieved from Bcipep and IEDB to evaluate the performance of the proposed prediction model. The experimental results showed that the proposed multi-expert system performed better than the six individual prediction system in general. The F1-Score of the proposed system could achieve 69.40% and 60.54% respectively, while the average F1-Scores of the other six systems could only achieve 55.21% and 41.94%. The proposed multi-expert recommendation system outperforms individual linear epitope prediction systems.

Keywords: Linear epitope · Multi-expert system · Polypeptide vaccine · Grouper · Vibrio infection

1 Introduction

Fishing industry plays an important role in economy for island countries. Located in the western Pacific Ocean, Taiwan is surrounded by ocean; hence, a lot of research efforts were focused on improving fisheries technologies. Fishery breeding and farming technologies were successfully achieved by Taiwan. Especially for aquarium and commercial fishes, such as grouper, trout, and Chinese sturgeon, the increasing rate of aquaculture products were far beyond products of wild catches. Therefore, to increase aquaculture products, higher density fishery farming becomes a trend and aquaculture environments become vulnerable to bacteria or viruses which would cause serious losses. To avoid

© The Author(s), under exclusive license to Springer Nature Switzerland AG 2022
M. S. Bansal et al. (Eds.): ICCABS 2021, LNBI 13254, pp. 51–62, 2022.
https://doi.org/10.1007/978-3-031-17531-2_5

significant losses, nervous necrosis virus (NNV), iridovirus, and bacterial infection [1] should be carefully prevented during lava farming stages [2]. Traditionally, antibiotics were frequently used to treat or prevent bacterial infection. However, long-term usage of antibiotics would cause drug resistance and cause difficulties of infection control. A recent report showed that high percentage of isolated vibrio evolved resistance to common antibiotics [3]. Therefore, in this report, we aimed to analyze pathogens' protein sequences directly and provide biologists and immunologists effective predictions for *in vitro* and *in vivo* vaccine experiments. Moreover, to strengthen immune systems of the fishes could decrease the usage of antibiotics and avoid antibiotic drug resistance [4].

In order to design vaccines, predicting epitope binding segments is indispensable pre-processes. Despite efforts that were made to develop linear epitopes (LE) predictors, yet recent technology can't provide outrageous predictive accuracy. According to previous reports of various LE predictors, the average value of the accuracy is only approximately 60% [5–10]. Intending to enhance the effectiveness of LE predictions and decrease experimental resources. We evaluated six popular LE predictors through various antigen protein sequences. With predicted results collected from these predictors, in order to propose a reliable prediction, we have investigated the superiority of each predictors for different sequences by using statistical analysis methods. Then, the integration of multiple predicting results was adopted to provide recommendation through a weighted voting mechanism for highly antigenic segments. Recommendation of effective segments for immunologists could be used to verify epitopes and to produce subunits and polypeptide vaccines.

2   Materials and Methods

2.1   Data Collection

In order to develop the Multiple Expert Linear Epitope Prediction System (MELEPS), we collected prediction results from multiple online tools. The MELEPS system considers six linear B-cell epitope prediction tools, including ABCpred [6], BCPREDS [7], BepiPred2.0 [8], LEPS [9], Bcepred [10], and LBtope [11]. To train the MELEPS system, we generated the Antigen protein Sequence Training (AST) dataset containing B-cell epitopes from the Immune Epitope Database (IEDB) [12, 13], which contained 2,628 protein sequences from NCBI Protein databases [14, 15] and a total of 26,945 experimentally validated antigenic epitopes with 399,876 amino acids were included. In order to verify the performance of MELEPS, we constructed two testing datasets. The Fishes Hosts (FH) dataset consisting 43 epitopes within 11 protein sequences for fish related infection was applied in this study. Since several prediction systems applied training data from Bcipep, a Bcipep Epitopes (BE) dataset for training was constructed from Bcipep [16]. The BE dataset was composed of 818 epitopes within 197 antigen proteins and retrieved from UniProt (Swiss-Prot) [17]. Distribution of twenty amino acids for all collected epitopes within the constructed three datasets were statistically analyzed and shown in Table 1. These datasets were used for training, measuring, and comparing the proposed MELEPS system.

Table 1. Quantity of each standard amino acid of epitopes in the three datasets

Type	Qty.	Type	Qty.	Type	Qty.	Type	Qty.	Type	Qty.
AST dataset									
A	28273	N	21626	H	8104	M	6818	T	24016
F	14524	E	28432	L	29369	P	25735	V	22199
C	6904	Q	19047	I	18767	R	19997	W	4540
D	25431	G	28557	K	25630	S	27981	Y	13926
FH dataset									
A	45	N	37	H	19	M	4	T	38
F	24	E	26	L	37	P	32	V	48
C	6	Q	18	I	25	R	34	W	11
D	29	G	43	K	27	S	42	Y	22
BE dataset									
A	504	N	339	H	151	M	115	T	435
F	226	E	501	L	481	P	475	V	427
C	143	Q	285	I	346	R	423	W	125
D	391	G	593	K	467	S	437	Y	243

2.2 The System Flow of MELEPS

The system architecture of the MELEPS included three main modules and was shown in Fig. 1. First, a target antigen sequence was sent into six predictors individually. The prediction results were analyzed through web scraper. All prediction results were integrated to calculated weighted or non-weighted recommendation scores. In the last module, low scoring candidate segments were filtered and a minimum continuous segment length was examined. Predicted results were displayed in the MELEPS output interface.

2.3 Integrated Multi-expert Recommendation Methodology

The multi-expert recommendation method combined training results from the six linear epitope prediction tools based on their default settings. Each linear epitope predictor was performed by different prediction methods. For instance, Bcepred was mainly based on physicochemical properties [9] and ABCpred based on recurrent neural network [6]. The constructed recommendation system explored different merits from various prediction methods to enhance broad properties for an overall recommendation result.

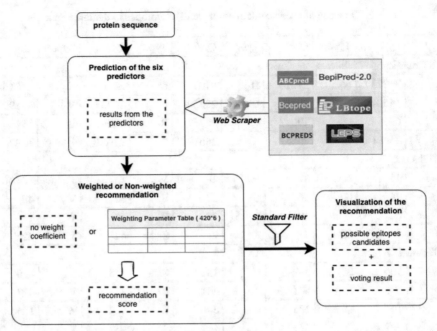

Fig. 1. The system architecture of the MELEPS. Module 1: a target antigen sequence was sent to the six predictors respectively. The prediction results were analyzed through web scraper. Subsequently, all prediction results were integrated to calculated weighted or non-weighted recommendation scores (step 2). Finally, for the last module, low scoring candidates were filtered and a continuous segment minimum length was examined (step 3). Predicted results were displayed in the MELEPS output page.

Weighted Recommendation. In this study, we assigned a corresponding weight to each predictor. The weight was calculated according to the performance of the predictor based on training dataset evaluation. To determine weighted voting score, we considered single amino acid (*AA*) and amino acid pair (*AAP*) as basic units. Hence, 20 amino acids and 400 possible paired combinations (20AAs * 20AAs) were considered as features for evaluating corresponding parameters, and a total of 420 corresponding weighting parameters for each expert prediction system were used to improve the reliability of multi-expert system. We used the antigen sequences in the AST dataset to perform statistical analysis on training results, and corresponding weight parameters of 420 parameters for each expert system were assigned. Afterward, these weighting parameters of MELEPS provide each expert system a corresponding reference look-up table that calculated a total score of predicted linear epitopes from a target antigen protein sequence according to the proportional and effective weighted recommendation.

Non-weighted Recommendation. The proposed MELEPS provides an alternative non-weighted recommendation approach that considered prediction results from the six linear epitope predictors without any bias weightings. Each predictor provides an identical weighting of 0.167 for all 420 features. In other words, each vote from an expert system holds an equivalent value for a target antigen protein prediction.

2.4 The Weighted Recommendation Score

To emphasize disparity of prediction accuracy for the six expert predictors, the weighted recommendation proposed a weighted voting mechanism. Every vote from different predictors on each amino acid or amino acid pair reflected inequivalent value in final recommendation. We defined weighting parameters for each expert system based on its prediction performance on training datasets. Weighted scores on each AA or AAP pair were calculated according to the following equation:

$$W_j = \sum_i S_i(AA_j) * \frac{1}{k}\left(P_i(AA_j) + P_i\left(AAP_{(j-1,j)}\right) + P_i\left(AAP_{(j,j+1)}\right)\right) (j = 2, \ldots, n-1) \tag{1}$$

where W_j defined as the weighted recommendation score of the jth residue in the protein sequence; AA_j defined as the jth amino acid in the sequence; S_i defined as the ith predictor; $S_i(AA_j)$ defined as the prediction result of jth residue in the target protein sequence from the ith predictor. The dynamical variable P_i represented the weighting parameter, which was the ratio of F1-score of the ith expert system to the sum of the F1-score from the six expert prediction systems for the single AA or AAP. In the equation, $P_i(AA_j)$ was the value of the weighting parameter for the jth AA in the sequence from the ith predictor; $FR_i\left(AAP_{(j-1,j)}\right)$ and $FR_i\left(AAP_{(j,j+1)}\right)$ were the scores of the AAP from the ith predictors, where $AAP_{(j-1,j)}$ represented the pair of residues composed of AA_{j-1} and AA_j, and $AAP_{(j,j+1)}$ was the pair of AA_j and AA_{j+1}. The constant k was defined as the number of P_i variables that were included. On one hand, when j was in the interval between 2 to n − 1, the constant $k = 3$. On the other hand, when $j = 1$ or $j = n$, only one of AAP exists; hence, $k = 2$. The P_i variable was calculated by the following equation:was normalized to the range of

$$P_i(R) = \frac{F1_i(R)}{\sum_i F1_i(R)} \tag{2}$$

where R was either AA or AAP; $F1_i(R)$ defined as the f1-score of R's from the ith predictor; $\sum_i F1_i(R)$ denoted the sum of the F1-score of R's of all predictors.

According to the equation mentioned above, the recommendation was based on the prediction results from the six linear epitope predictors respectively. When the residue was identified as positive epitope residue, the P_i weighting parameter was added as the corresponding weighted score. Otherwise, the parameter would be assigned to zeros. Finally, the P_i parameter was normalized to the range of [0, 1]. If a residue was predicted as positive epitopes by all six expert systems, the P_i parameters were assigned and the sum would be one. On the contrary condition, if prediction results of a residue were considered as negative epitope residue for six predictors simultaneously, the P_i parameter would be assigned to zero.

2.5 Performance Measurement of Recommendation

As stated above, every residue in the observed antigen sequence received a recommendation score which integrated the analysis results from six predictors. For either weighted or non-weighted recommendation methods, the recommendation scores were normalized to the range of [0, 1]. For each residue in a query antigen sequence, we checked whether the recommendation score met a specific threshold and decided the residue as an epitope or a non-epitope. Furthermore, we considered the minimum continuous length of an epitope peptide for the overall prediction. Eventually, the predicted epitope segments provided references for biologists and immunologists to design experiments on antigens.

To evaluate the performance of the MELEPS system, we calculated five indicators, including (1) sensitive (SEN) as the rate of true epitopes that were predicted correctly; (2) specificity (SPE) as the percentage of true non-epitopes that were predicted as non-epitopes correctly; (3) positive predictive value (PPV) as the percentage of positive prediction as true epitopes; (4) accuracy (ACC) as the rate of true epitopes and true non-epitopes were correctly predicted; (5) F1score as a combination measurement of both SEN and PPV into a single indicator within a range of [0, 1]. These indicators were used for comparing analysis results of the proposed MELEPS against the six individual predictors.

3 Results and Discussion

3.1 The Weight Parameter Table

The first performance measurements using the AST dataset to evaluate the six expert predictors. We calculated the defined six indicators on each AAs and AAPs for each predictor. Among the indicators, the F1-score was selected as the basis for weighting parameters.

We measured efficiency of the weighted parameter table based on F1-score by comparing with non-weighted recommendation method. Both BE and FH datasets were used to test the performance. The comparison results were shown in Table 2. As a consequence, for the FH dataset, the PPV of weighted method (71.96%) is higher than non-weighted approach (70.90%); for the BE dataset, the weighted method (86.73%) also performed a better PPV than the non-weighted method (85.68%). It is worthy to notice the indicator of PPV which is important factor for biologists during biological experiment design. Higher accuracy rates of PPV represent a better prevention to avoid false positive result occurrences. Hence, it could lower the cost and time consumption during verification experiments for vaccine development.

Table 2. Comparison between weighted and non-weighted recommendation method

Methods	ACC	SEN	SPE	PPV	F1-Score
FH dataset					
Weighted	52.95%	52.14%	54.76%	**71.96%**	60.46%
Non-weighted	53.79%	52.81%	55.75%	70.90%	60.54%
BE dataset					
Weighted	59.55%	56.19%	70.93%	**86.73%**	68.20%
Non-weighted	60.75%	57.17%	71.43%	85.68%	68.58%

3.2 Performance of MELEPS

The BE and the FH datasets were applied to estimate the performance of MELEPS (Fig. 2 and Fig. 3). According to previous report, the BE dataset comprised 818 epitopes that were originated from the UniProt (Swiss-Prot) database. The FH dataset focused on the epitopes which was observed from the antigens that infected fishes. The dataset contained 43 epitopes and 11 antigen protein sequences. These epitopes can be retrieved from the NCBI protein database. All epitopes were guaranteed as non-redundant segments. With these two datasets, we compared the prediction performant together.

As expected, MELEPS provides better performance in these two datasets than all other individual predictor. Table 3 shows that the weighted recommendation method in MELEPS with the best PPVs in the FH and BE datasets (72.94% and 89.84%). It indicates that MELEPS could reduce the possibility of false positive prediction and increase prediction efficiency. Furthermore, MELEPS shows better SPEs in both datasets (55.75% and 75.27%). Though it is a little bit lower than BCPREDS (76.52%) in BE dataset. However, the ACC of MELEPS consistently performed better than the average ACC of all the six predictors regarding both FH and BE datasets. According to the values of F1-score, the overall performance of MELEPS was also providing a stable performance than all the six existed predictors.

Table 3. Comparison of 5 indicators for different predictors.

System	SEN	SPE	ACC	PPV	F1score
FH dataset					
MELEPS_weighted	51.50	53.57	52.12	**72.49**	60.22
MELEPS_non-weighted	52.81	**55.75**	53.78	70.90	**60.54**
ABCPred	50.05	50.07	50.06	57.14	53.36
BcePred	50.12	50.05	50.07	31.39	38.61
BCPREDS	47.57	48.23	47.95	40.04	43.48
BepiPred2.0	51.27	51.45	51.35	54.67	52.92
LBtope	**59.43**	53.54	**55.15**	32.45	41.98
LEPS	57.26	50.93	51.65	13.05	21.26
six_Avg.	*52.62*	*50.71*	*51.04*	*38.12*	*41.94*
BE dataset					
MELEPS_weighted	56.53	75.27	60.38	**89.84**	69.40
MELEPS_non-weighted	57.17	71.44	60.75	85.68	68.58
ABCPred	50.54	50.82	50.65	61.21	55.36
BcePred	54.91	52.20	53.04	33.99	41.99
BCPREDS	63.84	**76.52**	**68.19**	83.90	**72.51**
BepiPred2.0	54.33	55.70	54.92	61.76	57.80
LBtope	61.08	54.39	56.29	34.68	44.24
LEPS	**78.25**	62.44	67.27	47.84	59.38
six_Avg.	*60.49*	*58.68*	*58.39*	*53.90*	*55.21*

To ensure effectiveness of the predicted epitopes, we analyzed prediction quality on a few selected antigen protein sequences in FH dataset. As shown in Table 4, MELEPS performed a better average F1score with a value of 59.32% than all other predictors. Taking the antigen protein of *R.sdmoninarum* as an example, it is a member of the *Micrococcaceae* family that causes bacterial kidney disease (BKD) in young salmonid fish. The *R.sdmoninarum* is an important bacterial pathogen of fish that has been widely discussed. The proposed system could predict epitope candidates with good F1-score of 74.37% and 74.80% regarding weighted and non-weighted approaches respectively. The comparison results with all other predicting systems were shown in Table 4.

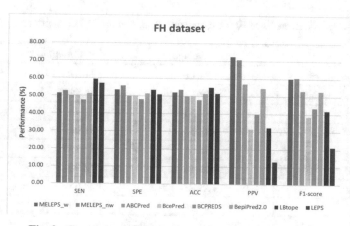

Fig. 2. Comparison of the seven predictors on the FH dataset.

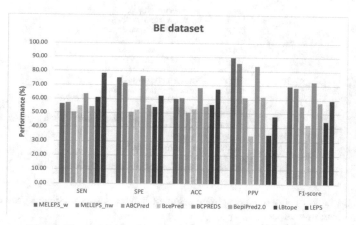

Fig. 3. Comparison of the seven predictors on the BE dataset.

In addition, we further examined the predicted epitopes through structural analysis. An antigen protein without resolved protein structure was firstly predicted by Phyre2 and I-TASSER for its corresponding virtual protein structure. The predicted epitope segments from our proposed system would be validated through their geometrical position on the 3-D structure. Since the true epitopes should locate on protein surface regions for binding activities with antibodies. For example, *Vibrio Brasiliensis DnaK* is considered as a candidate antigen protein that binds with cell receptors in a fish. In Fig. 4, we showed one of the predicted epitope segments (I_{203}EIDEVEGEKTFEVLATNGDT$_{223}$) lighted in red on the predicted protein structure. In this case, most of predicted epitope residues were located on the protein surface regions.

Table 4. Comparison results of F1score by using *R.sdmoninarum* as a query antigen protein.

System	Average	*R.sdmoninarum*
MELEPS_weighted	59.11	74.37
MELEPS_non-weighted	**59.32**	**74.80**
ABCPred	45.34	72.93
BcePred	37.57	54.97
BCPREDS	34.84	55.33
BepiPred2.0	49.44	72.98
LBtope	35.23	27.63
LEPS	21.64	39.26

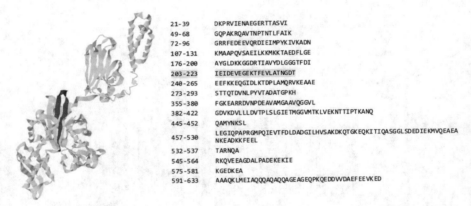

21-39	DKPRVIENAEGERTTASVI
49-68	GQPAKRQAVTNPTNTLFAIK
72-96	GRRFEDEEVQRDIEIMPYKIVKADN
107-131	KMAAPQVSAEILKKMKKTAEDFLGE
176-200	AYGLDKKGGDRTIAVYDLGGGTFDI
203-223	IEIDEVEGEKTFEVLATNGDT
240-265	EEFKKEQGIDLKTDPLAMQRVKEAAE
273-293	STTQTDVNLPYVTADATGPKH
355-380	FGKEARRDVNPDEAVAMGAAVQGGVL
382-422	GDVKDVLLLDVTPLSLGIETMGGVMTKLVEKNTTIPTKANQ
445-452	QAMYNKSL
457-530	LEGIQPAPRGMPQIEVTFDLDADGILHVSAKDKQTGKEQKITIQASGGLSDEDIEKMVQEAEA NKEADKKFEEL
532-537	TARNQA
545-564	RKQVEEAGDALPADEKEKIE
575-581	KGEDKEA
591-633	AAAQKLMEIAQQQAQAQQAGEAGEQPKQEDDVVDAEFEEVKED

Fig. 4. An example of a predicted epitope segment located on its predicted protein structure (*Vibrio Brasiliensis DnaK*).

3.3 The MELEPS Platform

A website was implemented to provide a user-friendly interface for biologists. The MELEPS accepts either single or multiple FASTA format sequence. The recommendation thresholding setting including minimum scores and length of an epitope could be set by users (Fig. 5). In addition, the users can choose either weighted or non-weighted recommendation method for the multi-expert prediction. Outcome of the voting mechanism on each amino acid within the query sequence and predicted epitope candidates would be displayed in the resulting pages. Figure 5 showed an example of DnaK predicted by MELEPS with the weighted recommendation method. The thresholding for each predictor were set as 0.2 with a minimum length of 1.

Fig. 5. MELEPS: query and resulting interfaces for protein DnaK.

4 Conclusion

In this study, we have considered six well-known linear epitope predictors and demonstrated their prediction abilities against different antigen datasets. It showed that each predictor adopted distinct training models and algorithms, and the performance are dynamically changed for various antigen protein sequences. To improve overall predictive effectiveness and stable performance, we proposed a multi-expert LE prediction system, MELEPS, which provided an integrated recommendation based on the six expert predictors. A weighted voting mechanism was implemented according to their individual performance regarding their training results on varied antigen protein datasets. By calculating the quality score of each predictor on each type of amino acid, a weighted score was assigned for weighting operation in the following prediction procedures. After comparing MELEPS with the original six predictors, MELEPS provided a higher accuracy and constantly stable performance, which proved substantial supports for subunit vaccine design. However, the best overall performance for the selected dataset could only reached 69.40% in F1-score, thus there is still room for improvement. It is mainly due to lack of balanced experimented linear epitopes. For the defined 420 features used in the weighted voting mechanism, the quantity of each features is imbalanced. We believed that the extension of several additional features could improve the effectiveness of the weighting parameters and break through the accuracy limit of epitope prediction.

Acknowledgments. This research was supported by the National Science Council, Taiwan (MOST 109-2321-B-019-005 to Prof. H.-Y. Chou, and MOST 110-2813-C-019-025-B to N.-S. Hwang).

References

1. Lai, W.Y., Xu, R.H., Chen, T.J., Wu, J.L.: Taiwan grouper research and development energy (R&D energy) and advantages. Agric. Biotechnol. Ind. Q. (26), 51–57 (2011). Taiwan Shibanyu Yanfa Nengliang yu Youshi
2. Shen, S.X., Zheng, A.C., Liu, B.Z., Lin, Z.H., Ran, F.H.: Status and trends of Taiwan's grouper industry. Agric. Biotechnol. Ind.Q. (38), 1–7 (2014). Taiwan Shibanyu Chanye Xiankuang yu Qushi
3. Amalina, N.Z., et al.: Prevalence, antimicrobial susceptibility and plasmid profiling of Vibrio spp. isolated from cultured groupers in Peninsular Malaysia. BMC Microbiol. **19**(1), 1–15 (2019)

4. WHO Vaccine Safety Basics. https://vaccine-safety-training.org/subunit-vaccines.html
5. Saha, S., Raghava, G.P.S.: Prediction of continuous B-cell epitopes in an antigen using recurrent neural network. Proteins Struct. Funct. Bioinform. **65**(1), 40–48 (2006)
6. Malik, A.A., Ojha, S.C., Schaduangrat, N., Nantasenamat, C.: ABCpred: a webserver for the discovery of acetyl-and butyryl-cholinesterase inhibitors. Mol. Diversity **26**, 1–21 (2021)
7. El-Manzalawy, Y., Dobbs, D., Honavar, V.: Predicting linear B-cell epitopes using string kernels. J. Mol. Recogn.: Interdisc. J. **21**(4), 243–255 (2008)
8. Jespersen, M.C., Peters, B., Nielsen, M., Marcatili, P.: BepiPred-2.0: improving sequence-based B-cell epitope prediction using conformational epitopes. Nucleic Acids Res. **45**(W1), W24–W29 (2017)
9. Wang, H.W., Lin, Y.C., Pai, T.W., Chang, H.T.: Prediction of B-cell linear epitopes with a combination of support vector machine classification and amino acid propensity identification. J. Biomed. Biotechnol. (2011)
10. Saha, S., Raghava, G.P.S.: BcePred: prediction of continuous B-cell epitopes in antigenic sequences using physico-chemical properties. In: Nicosia, G., Cutello, V., Bentley, P.J., Timmis, J. (eds.) Artificial Immune Systems, vol. 3239, pp. 197–204. Springer, Heidelberg (2004). https://doi.org/10.1007/978-3-540-30220-9_16
11. Singh, H., Ansari, H.R., Raghava, G.P.: Improved method for linear B-cell epitope prediction using antigen's primary sequence. PLoS ONE **8**(5), e62216 (2013)
12. IEDB. https://www.iedb.org/. Accessed 11 Oct 2021
13. Vita, R., et al.: The immune epitope database (IEDB): 2018 update. Nucleic Acids Res. (2018). PubMed PMID: 30357391, https://doi.org/10.1093/nar/gky1006
14. NCBI Protein database. https://www.ncbi.nlm.nih.gov/protein. Accessed 11 Oct 2021
15. NCBI. https://www.ncbi.nlm.nih.gov/. Accessed 11 Oct 2021
16. Saha, S., Bhasin, M., Raghava, G.P.: Bcipep: a database of B-cell epitopes. BMC Genomics **6**(1), 1–7 (2005)
17. UniProt: the universal protein knowledgebase in 2021. Nucleic Acids Research 49, no. D1, D480-D489 (2021)

Encoder-Decoder Architectures for Clinically Relevant Coronary Artery Segmentation

João Lourenço-Silva[1]([✉]) [iD], Miguel Nobre Menezes[2] [iD], Tiago Rodrigues[2], Beatriz Silva[2], Fausto J. Pinto[2] [iD], and Arlindo L. Oliveira[1] [iD]

[1] INESC-ID/Instituto Superior Técnico, University of Lisbon, Lisbon, Portugal
{joao.lourenco.silva,arlindo.oliveira}@tecnico.ulisboa.pt
[2] Cardiology Department, CAML, CCUL, Lisbon School of Medicine, Lisbon, Portugal

Abstract. Coronary X-ray angiography is a crucial clinical procedure for the diagnosis and treatment of coronary artery disease, which accounts for roughly 16% of global deaths every year. However, the images acquired in these procedures have low resolution and poor contrast, making lesion detection and assessment challenging. Accurate coronary artery segmentation not only helps mitigate these problems, but also allows the extraction of relevant anatomical features for further analysis by quantitative methods. Although automated segmentation of coronary arteries has been proposed before, previous approaches have used non-optimal segmentation criteria, leading to less useful results. Most methods either segment only the major vessel, discarding important information from the remaining ones, or segment the whole coronary tree, based mostly on contrast information, producing a noisy output that includes vessels that are not relevant for diagnosis. We adopt a better-suited clinical criterion and segment vessels according to their clinical relevance. Additionally, we simultaneously perform catheter segmentation, which may be useful for diagnosis due to the scale factor provided by the catheter's known diameter, and is a task that has not yet been performed with good results. To derive the optimal approach, we conducted an extensive comparative study of encoder-decoder architectures trained on a combination of focal loss and a variant of generalized dice loss. Based on the EfficientNet and the UNet++ architectures, we propose a line of efficient and high-performance segmentation models using a new decoder architecture, the EfficientUNet++, whose best-performing version achieves a generalized dice score of 0.9202 ± 0.0356, and artery and catheter class dice scores of 0.8858 ± 0.0461 and 0.7627 ± 0.1812.

1 Introduction

Coronary arteries are the blood vessels that carry oxygen and nutrient-rich blood to the heart tissue. Coronary Artery Disease (CAD), also known as Coronary Heart Disease or Ischemic Heart Disease, is a disease caused by the partial or

© The Author(s), under exclusive license to Springer Nature Switzerland AG 2022
M. S. Bansal et al. (Eds.): ICCABS 2021, LNBI 13254, pp. 63–78, 2022.
https://doi.org/10.1007/978-3-031-17531-2_6

complete blockage of the coronary arteries, which leads to limited or even ceased blood flow to the heart tissue and consequent myocardial dysfunction, being the cause of roughly 16% of global deaths every year [34].

X-ray coronary angiography (CAG) is one of the main procedures for CAD diagnosis and treatment. Traditionally, physicians use CAG images to assess the presence of stenosis, i.e., artery narrowing, through visual inspection. However, this method's subjectivity and potential unreliability led to the development of Quantitative Coronary Angiography (QCA), a diagnostic support tool that uses semi-automatic edge-detection algorithms to report vessel diameters at user-specified locations and the point of stenosis. Nevertheless, the low contrast and resolution of CAG images, the uneven contrast agent distribution, and the presence of artifacts, such as pacemakers, the spine, and the catheter itself, make this task very challenging. Thus, QCA still often requires manual correction of vessel boundaries. Furthermore, QCA only allows the analysis of a small vessel section at a time, limiting its use in clinical practice, in which the severity of stenosis is still assessed visually in most cases, rather than with QCA software.

Recently, deep learning methods have significantly improved coronary artery segmentation performance in CAG images, promising to overcome the faults of QCA's edge-detection algorithms. Most of them either segment only the major vessel [21,42,45,46] or try to segment the whole coronary tree, based primarily on contrast differences [8,35,49,52]. These criteria, however, may not be clinically optimal. The former discards potentially damaged vessels whose lesions may not be negligible, and the latter includes secondary vessels that may not be relevant for either diagnostic or therapeutic purposes, potentially distracting physicians from the important ones. We circumvent these shortcomings by adopting a better-suited clinical criterion developed in collaboration with expert cardiologists, in which a vessel is only segmented if it is 2 mm or wider at its origin. Since thinner vessels have higher risks of poor intervention outcomes, these are usually approached conservatively. Furthermore, the minimum diameter of commercially available revascularization devices is 2 mm [14,36]. Therefore, collateral vessels with diameters below 2 mm at their origin are generally deemed inadequate for revascularization and, when interpreting angiograms, physicians tend to ignore them.

With more complex lesion assessment and anatomical feature extraction in mind, we also segment the catheter, whose known diameter provides a scale factor that may help models determine vessel width and be important for diagnostic. To the best of our knowledge, simultaneous catheter and coronary artery segmentation in CAG images has only been performed in one previous work [39], reporting dice score coefficients (DSCs) far inferior to ours.

To determine the best architecture for this task, we conducted an extensive comparative study of existing encoders and decoders, which provided insights into the best architectural patterns for this and, presumably, other medical image segmentation problems. Based on our findings, we propose a new computationally efficient and high-performing decoder architecture, the EfficientUNet++. Combined with an EfficientNet-B5 [38] encoder, the EfficientUNet++ achieved a generalized dice score (GDS) of 0.9202 ± 0.0356, and DSCs of 0.8858 ± 0.0461 and 0.7627 ± 0.1812, for the artery and catheter classes, respectively.

Overall, the main contributions of this paper are as follows:

1. We propose a new and better clinically-suited criterion for catheter and coronary artery segmentation in CAG images, in which vessels are only labeled as such if they are deemed relevant for diagnostic and therapeutic purposes;
2. We perform an extensive quantitative and qualitative comparison of the performance of existing encoders and decoders, which may provide valuable insights for other medical image segmentation tasks;
3. Based on the findings of our study, we propose a line of models with the best performance-computation trade-offs, from which practitioners can choose according to the available hardware and clinical needs.

2   Related Work

2.1   Major Vessel Segmentation

Previous work has shown that major vessel segmentation can be improved by replacing the U-Net's encoder with popular image classification backbones, either pre-trained on ImageNet [45, 46] or trained from scratch on a relatively small dataset composed of 3200 CAG images [42]. Additionally, it has also been shown that using a modified generalized dice loss function with class weights to offset class imbalance and a tunable penalty for false positives and false negatives could further improve performance [45]. In the sequence of these findings, we train our models using a combination of the proposed loss function and Focal Loss (FL), and compare the performance of different state-of-the-art encoders.

Other authors have proposed a U-Net-based nested encoder-decoder architecture, the T-Net [21]. To simplify optimization, the authors replaced the U-Net's blocks with residual ones. In addition, to enable feature reuse, they arranged the pooling and up-sampling operations to make all the feature maps extracted by the encoder available to every layer of equal or greater depth of the decoder, in a DenseNet-like [17] fashion. These modifications, which enhance information flow through the network and enable it to outperform a standard U-Net, are also present in the UNet++ decoder [51] tested in this work.

2.2   Full Coronary Tree Segmentation

One of the main challenges of coronary artery segmentation is distinguishing vessels from artifacts. Since the former are only visible in the presence of contrast, previous work has used images acquired before contrast injection as a second-channel input to help a U-Net discern between vessels and background [8]. However, to be effective, this approach must be coupled with an image alignment algorithm that compensates for the motion caused by heartbeat and respiration. Furthermore, it requires the entire angiographic sequence to be acquired with minimal table motion, which can be hard to achieve, as standard clinical practice involves moving the patient table to follow the flow of dye within the vessels.

In line with what we propose in this paper, some authors have also attempted to use different decoder architectures to achieve better performance than what is possible with the commonly used U-Net. Specifically, they have used a pre-trained PSPNet [50,52], and a UNet++ combined with a feature pyramid network to improve multi-scale feature detection [49]. Other proposals include a deeply supervised encoder-decoder network with Gaussian convolutions, explicitly designed for vessel segmentation [35].

2.3   Catheter and Full Coronary Tree Segmentation

To the best of our knowledge, simultaneous catheter and coronary tree segmentation has only been addressed once [39]. Using a U-Net-based Siamese architecture trained on multi-class labels generated from low-level binary segmentation and optical flow, the authors obtained DSCs of 0.54 and 0.69 for the artery and catheter classes, respectively, far below the 0.8858 and 0.7627 DSCs we obtain.

2.4   Other Segmentation Criteria

As far as we know, an intermediate segmentation criterion, in which arteries with diameters inferior to 1 mm at their origin are not segmented, has only been proposed once [19]. However, the rationale behind this criterion is not explained, and it still includes many vessels that are ineligible for revascularization. To perform this task, the authors coupled a trainable preprocessing network with the U-Net and DeepLabV3+ architectures [3]. Even though their results demonstrate that the preprocessing module improves performance, that is not the focus of this work, and we leave the study of preprocessing methods for future work.

3   Model Evaluation Metrics

The segmentation quality of each class is measured using the DSC overlap metric. Let $\mathcal{G}$ and $\mathcal{P}$ be the sets of points belonging to a ground-truth segmentation mask and the segmentation mask predicted by a model, then, mathematically,

$$\text{DSC} = 2\frac{|\mathcal{G} \cap \mathcal{P}|}{|\mathcal{G}| + |\mathcal{P}|}. \tag{1}$$

Overall segmentation quality is measured using the GDS, another overlap metric. Let C be the number of classes, N the number of pixels, $g_{ci} \in \{0,1\}$ denote whether pixel i belongs to class c or not, and $p_{ci} \in [0,1]$ represent the probability of pixel i belonging to class c assigned by the model. Then,

$$\text{GDS} = 2\frac{\sum_{c=1}^{C} w_c \sum_{i=1}^{N} g_{ci}p_{ci}}{\sum_{c=1}^{C} w_c \sum_{i=1}^{N} g_{ci} + p_{ci}}, \tag{2}$$

where $w_c = 1/(\sum_{i=1}^{N} g_{ci})^2$ is the weight assigned to class c. These weights correct each class's contribution to the score by the inverse of its volume, reducing the correlation between region size and score [4]. Consequently, only models with good segmentation performance across all classes can achieve high GDS.

4   Loss Function

The problem we aim to solve can be interpreted as the combination of a macro-level and a micro-level one. The former consists of identifying the arteries and catheters, distinguishing them from each other and from artifacts, and determining which arteries have diameters equal or larger than 2 mm at their origin. The latter concerns the precise delineation of class contours, which are crucial to perform accurate anatomical measurements and reliable diagnoses.

To address these problems we trained the models on a loss function composed of a variant of GDL [37], named Penalty Generalized Dice Loss (pGDL) [45], which conveys information on global segmentation quality, and FL, which provides a pixel-wise evaluation focused on hard pixels, which are usually the ones that belong to less common classes and are near class boundaries and artifact regions. Using the notation defined in Sect. 3, the loss function can be defined as

$$\text{Loss} = \text{pGDL} + \lambda\text{FL} = 1 - \frac{\text{GDS}}{1 + k(1 - \text{GDS})} - \lambda\alpha(1 - p_{ci})^\gamma \log(p_{ci}), \quad (3)$$

where λ controls the relative weight of each term, k defines the magnitude of an additional penalty for false positives and false negatives, α balances the importance of positive and negative examples, and $(1 - p'_{ci})^\gamma$ is a modulating factor controlled by $\gamma \geq 0$ that down-weights easy examples and forces the model to focus on and learn from hard ones.

For simplicity, we use $\lambda = 1$ in all experiments. The hyperparameter k is set to $k = 0.75$, since we verified, in informal experiments, that this value works well for all models and leads to better performance than $k = 0$ for most. Following Lin et al. [28], we set the FL's hyperparameters to $\alpha = 0.25$ and $\gamma = 2$.

5   Architecture

Given the profusion of high-performing segmentation models, we started by conducting an extensive comparative study of existing encoders and decoders, aiming to determine the best models for this task and to identify the key components behind their performances. As a result, we propose a new computationally efficient and high-performing decoder architecture, the EfficientUNet++.

5.1   Encoder Comparison

Previous work [42, 45, 46] has shown that the U-Net's performance can be enhanced by replacing its encoder with more sophisticated backbones, both when pre-trained on a large dataset, like ImageNet [45, 46], and when trained from scratch on a relatively small dataset composed of 3200 CAG images [42].

Inspired by these findings, we tested multiple encoders on the coronary artery segmentation task. To avoid overfitting the encoders to our small dataset and evaluate the quality of the visual representations learned from ImageNet, we froze their weights during decoder training, which also shortened network training time and lowered GPU memory use during training. To investigate the existence of synergies between specific encoder-decoder pairs, we trained each backbone with multiple decoders: the U-Net [32], commonly used for medical image segmentation; the UNet++ [51], which has been shown to outperform the U-Net in multiple medical image segmentation tasks; and the DeepLabV3+ [3], a state-of-the-art semantic segmentation architecture. Figures 1a, 1b and 1c display segmentation performance as a function of FLOPS, when using encoders from the EfficientNet [38], RegNetY [31], ResNeXt [43] and ResNet [13] families.

Notably, for every decoder, the best performance is achieved using an EfficientNet backbone, suggesting that these models generalize and transfer better to new tasks. Furthermore, for the same performance, EfficientNet encoders are always more computationally efficient than other backbones. Due to their compound scaling, EfficientNet models are generally thinner at each scale than other encoders, i.e., use fewer channels to represent extracted features. Thus, since decoder computation scales with feature map dimension, EfficientNet backbones make for much more computationally efficient segmentation models than wider encoders. This is particularly evident when using complex decoders, such as the UNet++, that perform heavy processing on extracted features. Additionally, the efficient representations of EfficientNet encoders improve memory efficiency, allowing relatively larger training batches, which can be valuable when working with limited hardware resources.

The results indicate that, in general, better image classification architectures enable higher segmentation performance, which is in line with the widely accepted premise that there is a strong correlation between image classification performance and feature extraction capabilities. However, higher capacity encoders are not always better, and for some combinations of encoder family and decoder, performance starts degrading once the encoder exceeds a certain size. Since encoder weights are not updated during training and the same decoders converge to better solutions when combined with smaller encoder from the same family, the source of degradation appears to be decoder overfitting. This phenomenon is especially striking when using the UNet++, the decoder with more parameters and the one that applies more processing to encoder feature maps. Regularization techniques [24], a larger dataset, or both would probably mitigate or even avoid decoder overfitting. However, possibly due to their thin feature maps, EfficientNet encoders seem to have a regularizing effect on decoders too.

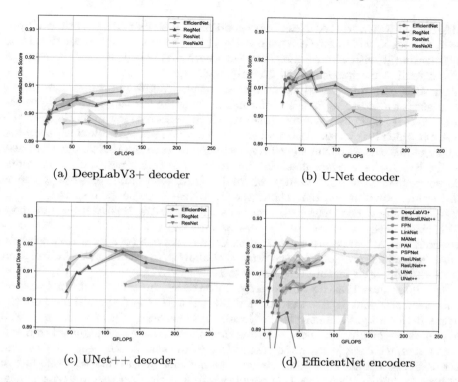

(a) DeepLabV3+ decoder

(b) U-Net decoder

(c) UNet++ decoder

(d) EfficientNet encoders

Fig. 1. Segmentation performance, measured by the GDS, as a function of FLOPS. Figures (a), (b) and (c) show the performance of different encoders combined with the (a) DeepLabV3+, (b) U-Net and (c) UNet++ decoders. Figure (d) shows the performance of different decoders combined with the EfficientNet B0 to B7 encoders. Each polygonal line corresponds to an encoder family. The markers represent the following models, in ascending order of FLOPS: EfficientNet - B0, B1, B2, B3, B4, B5, B6, B7; RegNet - Y2, Y4, Y6, Y8, Y16, Y32, Y40, Y64, Y80, Y120, Y160; ResNet - 18, 34, 50, 101, 152; ResNeXt - 50_32 × 4d, 101_32 × 4d, 101_32 × 8d. Above 250 GFLOPS, performance keeps degrading and is omitted. Models with GDS below 0.89 are also omitted.

5.2   Decoder Comparison

Given the superior performance and efficiency of EfficientNet backbones, each decoder was trained with multiple encoders from the EfficientNet family. Figure 1d shows the segmentation performance as a function of FLOPS for the DeepLabV3+ [3], FPN [23,27], PSPNet [50], U-Net [32] and the U-Net-based LinkNet [2], MA-Net [26], PAN [25], ResUNet [48], ResUNet++ [20] and UNet++ [51] decoder architectures.

The results denote the importance of the skip connections between encoder and decoder at every scale used by U-Net-based models. The UNet++ and the ResUNet++ achieve the best performance among all decoders, and the LinkNet,

MA-Net, U-Net, and ResUNet obtain good results, similar to each others'. However, the PAN, which builds on the U-Net, preserving the skip connections and augmenting it with attention mechanisms, performs poorly, suggesting the specific attention mechanisms used are prejudicial for this task.

In fact, the role played by attention mechanisms is not very clear. While they seem to harm the PAN's performance, they appear to be beneficial in the ResUNet++, and not affect the MA-Net, which performs very similarly to the U-Net, on which it is based. Also unclear is the importance of residual connections, which reduce the ResUNet's performance, compared to the U-Net, but work well in the ResUNet++. Since attention mechanisms and residual connections alone do not enhance the performances of the PAN, MA-Net, and ResUNet, it seems to be their combination that allows the ResUNet++ to perform so well.

Interestingly, the UNet++ performs similarly to the ResUNet++ but more efficiently, both parameter and computation-wise, without using residual connections nor attention. Instead, it uses densely connected nested decoder subnetworks, which promote feature reuse and allow it to extract more information from the encoder's feature maps at each scale.

Architectures that do not leverage localization information from low-level encoder feature maps, like the DeepLabV3+, the FPN and the PSPNet, are the worst-performing. While the lack of localization accuracy allows them to do well in generic segmentation tasks, it harms their performance when applied to medical images, which require fine segmentation. In the case of the DeepLabV3+ and the FPN, which use skip connections from encoder feature maps at one-fourth of original image resolution, this only leads to a slight decrease in performance, compared to U-Net-based models. On the other hand, for the PSPNet, which relies on localization information from encoder feature maps at one-eighth of input resolution, it results in very poor performance.

5.3   EfficientUNet++ Architecture

When coupled with EfficientNet backbones, the UNet++ achieves high segmentation performance with reasonable parameter and computational efficiency. However, while the number of parameters is not a major concern, the computation required for inference can be prohibitive of widespread clinical use, as it requires expensive hardware to be run promptly for entire angiographic sequences, usually comprised of about a hundred frames.

To address this, we propose a new architecture, the EfficientUNet++. Building on the UNet++, it reduces computational complexity by replacing its blocks with residual inverted bottleneck blocks with depthwise separable convolutions, and enhances performance by processing feature maps with concurrent spatial and channel squeeze and excitation (scSE) blocks [33], which combine the channel attention of squeeze and excitation (SE) blocks [16] with spatial attention.

As shown in Fig. 1d, when combined with EfficientNet encoders, the EfficientUNet++ establishes a line of efficient and high-performing segmentation models. Coupled with an EfficientNet-B5 encoder, it achieves an average GDS of 0.9202 and DSCs of 0.8858 and 0.7627 for the artery and catheter classes,

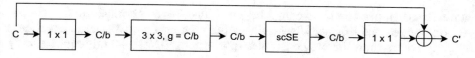

Fig. 2. EfficientUNet++'s convolutional block. Each convolution, except for the last, which is not activated, is followed by BN [18] and Hardswish activation [15]. C and C' are the numbers of input and output channels. Feature map height and width are not altered. We set the bottleneck ratio, b, to 1, and the number of convolution groups, g, to the number of input channels, making the 3×3 convolution depthwise. In the scSE block we use a squeeze ratio of 1.

respectively, outperforming all other models. Figures 3c and 3f display the segmentation masks produced by this model for a left coronary artery (LCA) and a right coronary artery (RCA), respectively.

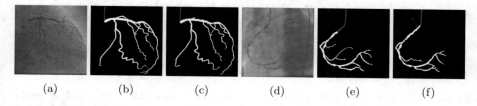

| (a) | (b) | (c) | (d) | (e) | (f) |

Fig. 3. Segmentation of (a-c) an LCA and (d-f) an RCA. Figures (b) and (e) corresponds to the ground-truth masks, and (c) and (f) are the masks produced by an EfficientNet-B5 encoder combined with an EfficientUNet++ decoder.

5.4 Performance vs. Computation Trade-Off

In clinical practice, depending on the available hardware resources and clinical needs, it may be necessary to make a trade-off between performance and computational efficiency. To help practitioners make that choice, we present, in Fig. 4, the Pareto frontier of all tested models, where each model is the most efficient at each performance level, and the best-performing at each computation regime. Thus, all other models need not be considered when looking for the best trade-off between performance and computational efficiency. Notably, neither the U-Net, UNet++ nor DeepLabV3+ appear in this plot, implying that they do not obtain the best trade-offs.

In Fig. 4, the lower a point is, the less computation it requires for inference, and the more to the left it is, the better its performance. Therefore, the gentler the slope between a model A and a better-performing model B, the more significant the relative merit of B compared to A. Considering that and the low performance of PSPNet decoders, the slightly more computationally demanding LinkNet-based architectures are probably the best choice when in the presence of a restrictive upper bound on computation. When less constrained by the computation budget, the EfiicientUNet++ decoder offers the best performance, being slightly better than the previously best-performing UNet++ and requiring only about a third of the computation (see Fig. 1d).

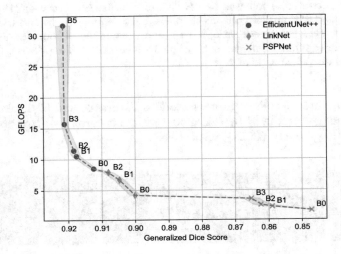

Fig. 4. FLOPS as a function of performance, measured by GDS. The dashed polygonal line corresponds to the Pareto frontier. Each marker represents a model: the text labels indicate the EfficientNet backbone, from B0 to B7, and the colors denote the decoder architecture.

6   Experimental Results

Figure 5 displays the GDS and DSC boxplots of the Pareto-efficient models determined in the previous section. Models using the PSPNet decoder are the worst-performing, only obtaining GDSs above 0.90 in 25% or less of the cases. LinkNet and EfficientUNet++ decoders obtain similar score distributions, with the latter achieve slightly larger mean and quartile values.

Despite the use of a loss function designed to handle class imbalance, artery segmentation performance is significantly superior to that of the lower volume catheter class. Also, being the most frequent class in the dataset, the artery class is the one that influences the GDS the most, and thus its DSC and the GDS follow similar trends. On the other hand, the DSC of the rarer catheter exhibits a more irregular behavior. For the EfficientUNet++ decoder, performance is high and consistent across cases, tending to improve when higher capacity encoders are used. However, for the remaining decoders, performance is rather inconsistent, and using larger scale encoders seems to harm performance instead of enhancing it, which may be an indicator of overfitting.

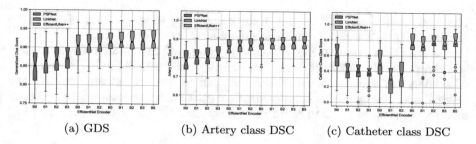

(a) GDS   (b) Artery class DSC   (c) Catheter class DSC

Fig. 5. Boxplot of the Pareto-efficient models' GDS and class DSCs. Each box corresponds to a model composed of an encoder denoted by the label in the x-axis and a decoder denoted by the color of the box. For each box, the line inside it and the cross represent the median and mean score of the respective model.

7   Implementation Details

7.1   Training Methodology

Encoders were pre-trained on ImageNet [6] and had their weights frozen during decoder training. Decoder parameters in hidden and output layers were initialized using Kaiming [12] and Xavier initialization [10], respectively. Each model was trained for 150 epochs using Adam [22] with $\beta_1 = 0.9$, $\beta_2 = 0.999$, a mini-batch size of 8, no weight decay, and an initial learning rate of 0.001, divided by ten at the 50th and 100th epochs. Most experiments were run using public PyTorch [30] implementations [44] under an MIT license. To keep comparisons fair, the repository was extended with ResUNet, ResUNet++ and EfficientUNet++ implementations. To obtain average scores and standard deviations, each model was trained and tested three times.

7.2   Dataset

Our dataset comprises 270 monochromatic 512×512 anonymized CAG images collected at a single center, between 2017 and 2019. The images were acquired from multiple viewing angles of LCA and RCA of 47 random patients, over the age of eighteen, who underwent CAG and invasive physiological assessment, i.e., fractional flow reserve (FFR), instantaneous wave-free ratio (iFR) or other indices' measurement. Each patient's angiograms were annotated by one of three expert cardiologists, blinded to the patient's identity, demographic information and medical history. Approximately one third of the patients underwent revascularization during or after the diagnostic angiography.

Measuring each vessel's diameter would have been impractical. Therefore, the catheter, whose diameter is known and varies between 1.8 mm and 2 mm, was used as a proxy to determine whether each vessel should be segmented or not. While this may have led to erroneous segmentation of some vessels with diameters close to 2 mm, it significantly simplified the annotation process. To diminish the quantity and impact of these and other errors, all annotations were reviewed multiple times by all three physicians.

We split the dataset at the patient level into a training, a validation and a test set, composed of 144/63, 21/9 and 23/10 images of the LCA/RCA, respectively. The images of each set were carefully chosen to keep it representative of the original one, having approximately identical distributions regarding the observed arteries, viewing angles and number of images annotated by each physician.

7.3   Data Augmentation

Our augmentation policy consists of the sequential application of the following random transformations: 1) $-20°$ to $20°$ rotation; 2) -10% to 10% horizontal and vertical translation; 3) -10% to 10% zoom; 4) -40% to 40% brightness variation, to account for brightness variability across devices. Following Fort et al. [9], we perform online augmentation and draw multiple augmentation samples per image in a growing batch regime. However, we keep the original images in the batch, as we observed this to improve performance slightly. Specifically, each batch is composed of two original images and three augmentations of each.

8   Discussion and Future Work

In this work, we propose a new and better clinically-suited criterion for catheter and artery segmentation in CAG images, developed in collaboration with expert cardiologists. Whereas most previous approaches either segment only the major vessel or the whole coronary tree, based mostly on contrast information, we only segment vessels relevant for diagnostic and therapeutic purposes.

To determine the best approach for the task, we conducted a comprehensive comparison of encoder and decoder architectures, whose results may prove useful for other medical image segmentation tasks. We found the EfficientNet [38] and the UNet++ [51] to be the best-performing encoder and decoder architectures, respectively. Due to their compound scaling, EfficientNet backbones are not only computationally and parameter efficient, but also in the way they represent features, generally using fewer channels at each scale than other models. This has a threefold benefit: 1) requires less computation from decoders; 2) seems to have a regularizing effect on decoders; 3) reduces memory use during training.

The performance of the UNet++ is related to its structure, whose densely connected nested decoders operating at different scales promote feature reuse and information flow through the network, and allow better multi-scale and overall processing of the features extracted by the encoder. Based on the UNet++, we propose a new computationally efficient and high-performing decoder architecture, the EfficientUNet++, which simultaneously increases its computationally efficiency and performance, through the use of light-weight depthwise separable convolutions and scSE spatial and channel attention blocks [33].

In the future, we plan to further improve our models through the use of self and semi-supervised learning techniques that allow us to take advantage of the tens of thousands of available unlabeled CAG images. Furthermore, we intend to test the EfficientUNet++ with attention mechanisms other than scSE, such as the CBAM [40] and Triplet Attention [29], and explore new architectures, such as Vision Transformers [7], whose features contain explicit semantic segmentation information when self-trained [1], and hybrid models combining convolutions and self-attention [5,11,41,47], which possess long-range modeling capabilities that may be crucial in the coronary artery segmentation task. Finally, we plan to investigate the clinical usefulness of our models, which is not necessarily determined by the DSCs and GDS they obtain.

Acknowledgements. This work was supported by national funds through Fundação para a Ciência e Tecnologia (FCT), under the project with reference UIDB/50021/2020.

References

1. Caron, M., et al.: Emerging properties in self-supervised vision transformers. arXiv preprint arXiv:2104.14294 (2021)
2. Chaurasia, A., Culurciello, E.: Linknet: exploiting encoder representations for efficient semantic segmentation. In: 2017 IEEE Visual Communications and Image Processing (VCIP), pp. 1–4. IEEE (2017)
3. Chen, L.C., Zhu, Y., Papandreou, G., Schroff, F., Adam, H.: Encoder-decoder with atrous separable convolution for semantic image segmentation. In: Proceedings of the European conference on computer vision (ECCV), pp. 801–818 (2018)
4. Crum, W.R., Camara, O., Hill, D.L.: Generalized overlap measures for evaluation and validation in medical image analysis. IEEE Trans. Med. Imaging **25**(11), 1451–1461 (2006)
5. Dai, Z., Liu, H., Le, Q.V., Tan, M.: Coatnet: marrying convolution and attention for all data sizes. arXiv preprint arXiv:2106.04803 (2021)
6. Deng, J., Dong, W., Socher, R., Li, L.J., Li, K., Fei-Fei, L.: Imagenet: a large-scale hierarchical image database. In: 2009 IEEE Conference on Computer Vision and Pattern Recognition, pp. 248–255. IEEE (2009)
7. Dosovitskiy, A., et al.: An image is worth 16x16 words: transformers for image recognition at scale. arXiv preprint arXiv:2010.11929 (2020)
8. Fan, J., et al.: Multichannel fully convolutional network for coronary artery segmentation in x-ray angiograms. IEEE Access **6**, 44635–44643 (2018)
9. Fort, S., Brock, A., Pascanu, R., De, S., Smith, S.L.: Drawing multiple augmentation samples per image during training efficiently decreases test error. arXiv preprint arXiv:2105.13343 (2021)
10. Glorot, X., Bengio, Y.: Understanding the difficulty of training deep feedforward neural networks. In: Proceedings of the Thirteenth International Conference on Artificial Intelligence and Statistics, pp. 249–256. JMLR Workshop and Conference Proceedings (2010)
11. Graham, B., et al.: Levit: a vision transformer in convnet's clothing for faster inference. arXiv preprint arXiv:2104.01136 (2021)
12. He, K., Zhang, X., Ren, S., Sun, J.: Delving deep into rectifiers: surpassing human-level performance on imagenet classification. In: Proceedings of the IEEE International Conference on Computer Vision, pp. 1026–1034 (2015)

13. He, K., Zhang, X., Ren, S., Sun, J.: Deep residual learning for image recognition. In: Proceedings of the IEEE Conference on Computer Vision and Pattern Recognition, pp. 770–778 (2016)
14. van der Heijden, L.C., et al.: Small-vessel treatment with contemporary newer-generation drug-eluting coronary stents in all-comers: insights from 2-year dutch peers (twente ii) randomized trial. Am. Heart J. **176**, 28–35 (2016)
15. Howard, A., et al.: Searching for mobilenetv3. In: Proceedings of the IEEE/CVF International Conference on Computer Vision, pp. 1314–1324 (2019)
16. Hu, J., Shen, L., Sun, G.: Squeeze-and-excitation networks. In: Proceedings of the IEEE Conference on Computer Vision and Pattern Recognition, pp. 7132–7141 (2018)
17. Huang, G., Liu, Z., van der Maaten, L., Weinberger, K.Q.: Densely connected convolutional networks. In: Proceedings of the IEEE Conference on Computer Vision and Pattern Recognition, pp. 4700–4708 (2017)
18. Ioffe, S., Szegedy, C.: Batch normalization: accelerating deep network training by reducing internal covariate shift. In: International Conference on Machine Learning, pp. 448–456. PMLR (2015)
19. Iyer, K., et al.: Angionet: a convolutional neural network for vessel segmentation in x-ray angiography. medRxiv (2021)
20. Jha, D., et al.: Resunet++: an advanced architecture for medical image segmentation. In: 2019 IEEE International Symposium on Multimedia (ISM), pp. 225–2255. IEEE (2019)
21. Jun, T.J., Kweon, J., Kim, Y.H., Kim, D.: T-net: nested encoder-decoder architecture for the main vessel segmentation in coronary angiography. Neural Netw. **128**, 216–233 (2020)
22. Kingma, D.P., Ba, J.: Adam: a method for stochastic optimization. arXiv preprint arXiv:1412.6980 (2014)
23. Kirillov, A., He, K., Girshick, R., Dollár, P.: Iccv_stuff_fair_final. http://presentations.cocodataset.org/COCO17-Stuff-FAIR.pdf. Accessed on 10 June 2021
24. Kukačka, J., Golkov, V., Cremers, D.: Regularization for deep learning: a taxonomy. arXiv preprint arXiv:1710.10686 (2017)
25. Li, H., Xiong, P., An, J., Wang, L.: Pyramid attention network for semantic segmentation. arXiv preprint arXiv:1805.10180 (2018)
26. Li, R., Zheng, S., Duan, C., Zhang, C., Su, J., Atkinson, P.: Multi-attention-network for semantic segmentation of fine resolution remote sensing images. arXiv preprint arXiv:2009.02130 (2020)
27. Lin, T.Y., Dollár, P., Girshick, R., He, K., Hariharan, B., Belongie, S.: Feature pyramid networks for object detection. In: Proceedings of the IEEE Conference on Computer Vision and Pattern Recognition, pp. 2117–2125 (2017)
28. Lin, T.Y., Goyal, P., Girshick, R., He, K., Dollár, P.: Focal loss for dense object detection. In: Proceedings of the IEEE International Conference on Computer Vision, pp. 2980–2988 (2017)
29. Misra, D., Nalamada, T., Arasanipalai, A.U., Hou, Q.: Rotate to attend: convolutional triplet attention module. In: Proceedings of the IEEE/CVF Winter Conference on Applications of Computer Vision, pp. 3139–3148 (2021)
30. Paszke, A., et al.: Pytorch: an imperative style, high-performance deep learning library. In: Wallach, H., Larochelle, H., Beygelzimer, A., d'Alché-Buc, F., Fox, E., Garnett, R. (eds.) Advances in Neural Information Processing Systems 32, pp. 8024–8035. Curran Associates, Inc. (2019). http://papers.neurips.cc/paper/9015-pytorch-an-imperative-style-high-performance-deep-learning-library.pdf

31. Radosavovic, I., Kosaraju, R.P., Girshick, R., He, K., Dollár, P.: Designing network design spaces. In: Proceedings of the IEEE/CVF Conference on Computer Vision and Pattern Recognition, pp. 10428–10436 (2020)
32. Ronneberger, O., Fischer, P., Brox, T.: U-Net: convolutional networks for biomedical image segmentation. In: Navab, N., Hornegger, J., Wells, W.M., Frangi, A.F. (eds.) MICCAI 2015. LNCS, vol. 9351, pp. 234–241. Springer, Cham (2015). https://doi.org/10.1007/978-3-319-24574-4_28
33. Roy, A.G., Navab, N., Wachinger, C.: Concurrent spatial and channel 'squeeze & excitation' in fully convolutional networks. In: Frangi, A.F., Schnabel, J.A., Davatzikos, C., Alberola-López, C., Fichtinger, G. (eds.) MICCAI 2018. LNCS, vol. 11070, pp. 421–429. Springer, Cham (2018). https://doi.org/10.1007/978-3-030-00928-1_48
34. Rudd, K.E., et al.: Global, regional, and national sepsis incidence and mortality, 1990–2017: analysis for the global burden of disease study. Lancet **395**(10219), 200–211 (2020)
35. Samuel, P.M., Veeramalai, T.: VSSC net: vessel specific skip chain convolutional network for blood vessel segmentation. Comput. Methods Programs Biomed. **198**, 105769 (2021)
36. Sim, H.W., et al.: Treatment of very small de novo coronary artery disease with 2.0 mm drug-coated balloons showed 1-year clinical outcome comparable with 2.0 mm drug-eluting stents. J. Invasive Cardiol. **30**(7), 256–261 (2018)
37. Sudre, C.H., Li, W., Vercauteren, T., Ourselin, S., Jorge Cardoso, M.: Generalised dice overlap as a deep learning loss function for highly unbalanced segmentations. In: Cardoso, M.J., et al. (eds.) DLMIA/ML-CDS -2017. LNCS, vol. 10553, pp. 240–248. Springer, Cham (2017). https://doi.org/10.1007/978-3-319-67558-9_28
38. Tan, M., Le, Q.: Efficientnet: rethinking model scaling for convolutional neural networks. In: International Conference on Machine Learning, pp. 6105–6114. PMLR (2019)
39. Vlontzos, A., Mikolajczyk, K.: Deep segmentation and registration in x-ray angiography video. arXiv preprint arXiv:1805.06406 (2018)
40. Woo, S., Park, J., Lee, J.Y., Kweon, I.S.: Cbam: convolutional block attention module. In: Proceedings of the European Conference on Computer Vision (ECCV), pp. 3–19 (2018)
41. Wu, H., et al.: CVT: introducing convolutions to vision transformers. arXiv preprint arXiv:2103.15808 (2021)
42. Xian, Z., Wang, X., Yan, S., Yang, D., Chen, J., Peng, C.: Main coronary vessel segmentation using deep learning in smart medical. Math. Prob. Eng. **2020** (2020)
43. Xie, S., Girshick, R., Dollár, P., Tu, Z., He, K.: Aggregated residual transformations for deep neural networks. In: Proceedings of the IEEE Conference on Computer Vision and Pattern Recognition, pp. 1492–1500 (2017)
44. Yakubovskiy, P.: Segmentation models pytorch. https://github.com/qubvel/segmentation_models.pytorch (2020)
45. Yang, S., Kweon, J., Kim, Y.H.: Major vessel segmentation on x-ray coronary angiography using deep networks with a novel penalty loss function. In: International Conference on Medical Imaging with Deep Learning-Extended Abstract Track (2019)
46. Yang, S., et al.: Deep learning segmentation of major vessels in x-ray coronary angiography. Sci. Rep. **9**(1), 1–11 (2019)
47. Yuan, K., Guo, S., Liu, Z., Zhou, A., Yu, F., Wu, W.: Incorporating convolution designs into visual transformers. arXiv preprint arXiv:2103.11816 (2021)

48. Zhang, Z., Liu, Q., Wang, Y.: Road extraction by deep residual u-net. IEEE Geosci. Remote Sens. Lett. **15**(5), 749–753 (2018)
49. Zhao, C., et al.: Semantic segmentation to extract coronary arteries in fluoroscopy angiograms. medRxiv (2020)
50. Zhao, H., Shi, J., Qi, X., Wang, X., Jia, J.: Pyramid scene parsing network. In: Proceedings of the IEEE Conference on Computer Vision and Pattern Recognition, pp. 2881–2890 (2017)
51. Zhou, Z., Rahman Siddiquee, M.M., Tajbakhsh, N., Liang, J.: UNet++: a nested U-Net architecture for medical image segmentation. In: Stoyanov, D., et al. (eds.) DLMIA/ML-CDS -2018. LNCS, vol. 11045, pp. 3–11. Springer, Cham (2018). https://doi.org/10.1007/978-3-030-00889-5_1
52. Zhu, X., Cheng, Z., Wang, S., Chen, X., Lu, G.: Coronary angiography image segmentation based on PSPnet. Comput. Methods Programs Biomed. **200**, 105897 (2021)

Unified SAT-Solving for Hard Problems of Phylogenetic Network Construction

Dan Gusfield[1]([⊠]) [iD] and Hannah Brown[1,2] [iD]

[1] Computer Science Department, University of California, Davis, USA
gusfield@cs.ucdavis.edu
[2] Computer Science Department, NUS, Singapore, Singapore

Abstract. We developed a unified SAT-solving approach to three phylogenetic network construction problems that integer linear programming was not able to solve. We detail the CNF formulations for two: the Minimum-Reticulation Problem, and the Hybridization-Network Problem; and apply them to widely studied Grass data, and to randomly generated data. The results reduce the number of reticulations in 4 of the 18 Grass datasets, and improve the best SAT-based solutions in two more. We study the effectiveness of a method for speeding the solution to the Hybridization problem, and show that when it can be applied, it can be extremely effective. This work contributes to SAT-solving methodology, to phylogenetic network construction, and to the evolutionary history of Grasses.

1 Introduction: Evolutionary Trees and Phylogenetic Networks

Trees are the traditional way to think about and represent evolution, due to evolutionary splitting (branching), but trees are often too simplistic, both for biological and methodological reasons. Instead, there are now many evolutionary problems that are framed in terms of finding **networks**, usually directed acyclic graphs (DAGs), that represent more complex evolution than do trees.

However, trees still carry considerable evolutionary information, and networks are often required to *contain* known trees; or to reflect partial information about trees. This motivated two major algorithmic problems that relate trees and networks, in addition to one more restricted problem. These problems are the *Minimum Reticulation Problem*; the *Hybridization-Network Problem*; and the *Hard-Wired Cluster-Network Problem* [8]. Due to limited space, we will only discuss the first two of these problems.

Previous attempts to solve practical instances of these problems using *integer linear programming* (ILP) were very disappointing [10].[1] This is in contrast to the *many* problems in Computational Biology where ILP has been very successful

---

[1] However, ILP was successful for a special case of one of the problems.

Thanks to NSF for supporting this research under the grant 1528234. Support for H.B. was from NSF Research Experiences for Undergraduates grant 1528234.

M. S. Bansal et al. (Eds.): ICCABS 2021, LNBI 13254, pp. 79–91, 2022.
https://doi.org/10.1007/978-3-031-17531-2_7

[6]. So, following the success of SAT-solving reported in [4], we next developed a *unified SAT-solving* approach to these three phylogenetic network problems.

In this paper, we explain that *unified* SAT approach, and discuss empirical results using both real and simulated data. We apply the methods to a well-studied dataset of grasses, confirming previous results in most cases, and reducing the size of needed networks in several cases. We also study the effectiveness of a *heuristic* to speed up the computation of the Hybridization problem, first developed for two trees [1–3], and later applied to multiple trees [13]. The paper contributes to SAT-solving methodology, to phylogenetic network construction, and to the evolutionary history of Grasses.

2 Definitions and Problem Statements

The Central Definition

Let S be a set of n *leaf* labels. These are the *taxa* of S, and each element in S is a *taxon*. A *Directed Acyclic Graph (DAG)*, $\mathcal{D}$, with n leaves labeled by the taxa of S, **displays** a *strict* subset s of S (called a *cluster* of S) if:

A. There is some tree T_s, with root node 0, embedded in $\mathcal{D}$, that reaches *all* of the leaves of $\mathcal{D}$. Tree T_s is called a *leaf-spanning* tree for s.

B. There is a node v in T_s, such that the leaves of T_s reachable from v have *exactly* the labels in subset s. That is, those labels are the *taxa* in subset s.

These two requirements are called *Display-Requirements*. The nodes and edges in T_s that are reachable from node v, as in Display-Requirement B, form a directed *subtree* of T_s, called a *cluster-tree* for s. See Fig. 1.

Reticulation Networks

Let F be a **family** of specified *clusters* of S. A DAG that displays *every* cluster in F is called a **Reticulation Network** for (S, F).[2] Any node with in-degree more than one is called a *reticulation node*. For example, let $S = \{a, b, c, d, e\}$, and $F = \{a, b\}, \{b, c\}, \{d, e\}, \{b, c, d, e\}$. DAG $\mathcal{D}$ (in Fig. 1) displays each of the clusters.

Minimum-Reticulation

Reticulation events in phylogenetics are generally due to *species hybridization*, or *lateral gene transfer*. In shorter time spans, reticulation can be due to *meiotic recombination* [5]. In all of these biological events, it is extremely rare to have a reticulation reflect the mixture of more than two ancestral lines. Hence, a reticulation node is generally restricted to have in-degree of *two*. Further, reticulation events are generally infrequent, so it is widely believed that reticulation networks with *fewer* reticulations more realistically *represent* true evolutionary history. These two biological assumptions lead to

The **Minimum-Reticulation Problem**: Given S and F, construct a reticulation network for (S, F), with the *fewest* reticulation nodes, t^*, over

---

[2] Such a network is called a *Soft-Wired Cluster Network* in [8].

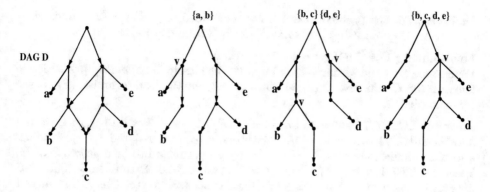

Fig. 1. DAG $\mathcal{D}$ and three *leaf-spanning* trees of $\mathcal{D}$. The two end trees each contain one cluster-tree for the cluster written above the leaf-spanning tree. The middle tree contains two cluster-trees, one for each of the displayed clusters in F.

all reticulation networks for (S, F), where every reticulation node has in-degree two.

Number t^* is also called the *hybridization number* for (S, F). If the in-degree is *not* bounded by two, the minimum number of reticulation nodes in any reticulation network for (S, F) is called the *reticulation number* for (S, F).

Previously, J. Matsieva and D. Gusfield, developed three integer linear programming (ILP) formulations for the Minimum-Reticulation problem, but (contrary to the successes of ILP for many problems in computational biology [6]) *none* of these ILP formulations could solve anything more than trivial problem instances. A top-down branching program was also developed by J. Matsieva [10] to construct a reticulation network for input (S, F). In the special case of *tree-child networks* the method is guaranteed to produce optimal networks. Before the work presented here, there was no available software that solves the Minimum Reticulation Problem in practice, for data of biological interest, motivatating our efforts to explore a SAT-solving approach.

The Target Feasibility Problem (TFP). The general approach to the Minimum-Reticulation Problem for an input (S, F), is to formulate the *subproblem* of determining, for a given *target* t, whether or not there is a reticulation network, $\mathcal{D}$, for input (S, F), that uses at most t reticulation nodes. This is called the **Target Feasibility Problem (TFP)**. Then, the Minimum-Reticulation Problem can be solved by starting with a value of t that is guaranteed to be larger or equal to (the unknown) t^*; and successively solving the *TFP*, reducing the target t by one each time, until target t is no longer attainable, at which point, $t^* = t + 1$.[3] The initial target t is given by an *upper bound* on the number of *recombination* nodes needed in any *Ancestral Recombination Graph (ARG)*

---

[3] Binary search is natural but not advised, because the solution time needed for an *UNsatisfiable* formula is generally *much larger* than for a *satisfiable* formula.

for (S, F). For the problem instances we examined, a good upper bound (close to t^*) can be found using programs *SHRUB* [12], and *KWARG* [9].

Formulating the TFP as a SAT problem

We assume that the reader is familiar with the classic *Satisfiability (SAT) Problem* and the *Conjunctive Normal Form (CNF)* for Boolean formulas. If not, see for example [11].

Super-DAG. H To solve an instance (S, F) of the TFP with target t, we use a *super-DAG*, H, which is a DAG with one root, that is large enough that an optimal reticulation network, $\mathcal{D}$, for (S, F) is guaranteed to be a *subgraph* of H. Then, the TFP is solved by selecting nodes and edges of H to form $\mathcal{D}$. The CNF formulas must ensure that $\mathcal{D}$ is a reticulation network for (S, F), with at most t reticulations.

It is fairly easy to prove that for $|S| = n$, *if* $\mathcal{D}$ needs at most t reticulation nodes, then it needs at most $n - 2 + 2t$ *interior* (non-leaf) nodes. So, H has $q = 2n - 2 + 2t + 1$ nodes numbered 0 through $q - 1$. Node 0 is the single *root* with a directed edge to each of the interior nodes; the last n numbered nodes are the leaves of H, with no out-edges; and each of the interior nodes have a directed edge *to* every node in H with a higher numbered label (including the leaves). See Fig. 2.

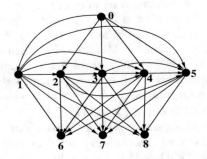

Fig. 2. Super-DAG H for $n = 3$ and $t = 2$, so it has $3 - 2 + 2 \times 2 = 5$ internal nodes, 1 through 5.

A feasible solution (if there is one) to an instance of the TFP, with target t, is a *subset* of nodes and edges of H. We find a solution (if there is one) by formulating the TFP instance as a CNF formula $\mathcal{F}$, and then use a SAT-solver to find a *satisfying* assignment of the variables of $\mathcal{F}$, or determine that $\mathcal{F}$ is *unsatisfiable*.

3   The CNF Formula for the TFP

The CNF formula for the TFP has three high-level components. The first two encode the A and B display-requirements (in the Central Definition); the third component, the most complex, contains the CNF clauses to determine whether a node in $\mathcal{D}$ is a reticulation node, and to restrict the number of reticulations to be at most the target t. The full CNF for an instance of the TFP has over forty-five types of clauses. We will describe the most important ones.

CNF Formulation for Display-Requirement A:
For each cluster s in F and each directed edge (i, j) in H, create the binary variable $X(s, i, j)$ whose value indicates whether edge (i, j) will be part of the constructed leaf-spanning tree T_s for cluster s. Similarly, for each cluster s in F and each node $j \neq 0$ in H, create the variable $T_{in}(s, j)$ to indicate whether there is *some* edge directed *into* node j of T_s. Then for each s and each node $j \neq 0$ in H, we have the CNF clause:

$$(X(s, i_1, j) \lor X(s, i_2, j) \lor ... \lor X(s, i_k, j)) \lor (\neg T_{in}(s, j)),$$

where the edges $(i_1, j)...(i_k, j)$ are all the edges into node j in H. This implements the logic that variable $T_{in}(s, j)$ can be set *true*, *only if* there is some variable $X(s, i, j)$ that is set *true*.

We also have a CNF clause: $(\neg X(s, i, j) \lor T_{in}(s, j))$, for *each* edge (i, j) in H that is directed into node j. These clauses implement the logic that $T_{in}(s, j)$ *must* be set *true if* there is some variable $X(s, i, j)$ that is set *true*, i.e., some edge *into* node j in T_s. Also, for each s and each node $j \neq 0$, and each *pair* of directed edges $(i, j), (i', j)$ into node j in H, we have the CNF clause: $\neg X(s, i, j) \lor \neg X(s, i', j)$, which together implement the logic that there can be *at most one* edge directed into node j in T_s.

Next, we have the single-variable clause $T_{in}(s, j)$, for each *leaf* j in H, which implements the requirement that there must be an edge into leaf j in the leaf-spanning tree created for cluster s, since every leaf of H must be in each leaf-spanning tree. Finally, for each cluster s, and each node $j \neq 0$, we have a CNF clause for the logic that if there is some (necessarily only one) edge into *node j* in T_s, and it is edge (i, j), then there must also be some edge in T_s *into* node i. That is expressed by the Boolean logic $T_{in}(s, j) \land X(s, i, j) \Rightarrow T_{in}(s, i)$, which is written in CNF as the clause: $(T_{in}(s, i) \lor \neg T_{in}(s, j) \lor \neg X(s, i, j))$.[4]

The above CNF clauses (ANDed together) have the combined effect that for each cluster s in F, the variables that are set to *true*, specify a subset of nodes and edges of H forming a leaf-spanning tree T_s. Essentially, we think of each tree T_s as growing from the leaves of H *upward* to the root of H. The CNF ensures that each leaf of H is in each T_s, and for each node $j \neq 0$ in each T_s, there must be a single directed edge (i, j) in H that is in T_s; and when edge (i, j) is in T_s, node i is also in T_s. These clauses ensure that each T_s is a leaf-spanning tree for cluster s.

The CNF Formulation for Display-Requirement B:
We use one variable $CT_s(i)$ indicating whether node i will be in the *cluster-tree* for cluster s; and one binary variable $RCT_s(i)$, for each node $i \neq 0$ to

---

[4] Recall that the relation $P \Rightarrow Q$ is equivalent to $\neg P \lor Q$; and that $\neg (P \land Q)$ is equivalent to $(\neg P \lor \neg Q)$, so $(P \land W) \Rightarrow Q$ is equivalent to $Q \lor \neg P \lor \neg W$.

indicate whether node i will be the *root* of that cluster-tree for s. Then, Display-Requirement B is implemented in CNF by a set of *forward* and *backward* clauses.

Forward Clauses for Requirement B: For each node $i \neq 0$, and each cluster s, we have binary variables $CT_s(i)$ and $RCT_s(i)$, which indicate, respectively, whether or not node i will be in the cluster-tree for s; and whether or not it will be the root of that cluster tree. Then the formulation will have the CNF clauses: $\neg RCT_s(i) \vee T_s(i)$, and $\neg RCT_s(i) \vee CT_s(i)$. These two clauses mean that node i can be chosen as the root of the cluster-tree for s, *only if i* is a node in the leaf-spanning tree T_s, and is a node in the cluster-tree for s.

Next, we need clauses to ensure that for each cluster s, *exactly* one node in T_s is selected to be the root of the cluster-tree for s. For any specific cluster s, we OR together, into one clause, the variables $RCT_s(i)$ for all the interior nodes in H. Then, for each cluster s and each pair of interior nodes $i \neq 0$ and $j \neq 0$, we use the clause $\neg RCT_s(i) \vee \neg RCT_s(j)$, to ensure that *at most* one root is selected. Many of these clauses are unnecessary, because some of the nodes in them are not in T_s, but this does not affect the correctness of the CNF formulation. Also, if j is a leaf of H, we have the single-variable clause $\neg RCT_s(j)$, which forbids j from being the root of the cluster tree.

The Most Important Relations: For a specific cluster s, any interior node i, and any edge (i, j) in H, we need the logic: If node i is in the cluster-tree for s, and there is a directed edge (i, j) in T_s, then node j must be in the cluster-tree for s. That is, $(CT_s(i) \wedge X_s(i, j)) \Rightarrow CT_s(j)$, which is implemented as: $\neg CT_s(i) \vee \neg X_s(i, j) \vee CT_s(j)$

Finally, for a leaf j labeled by a taxon in s, we have the single-variable clause $CT_s(j)$, which specifies that we want leaf j to be in the cluster-tree for s; and for a leaf j *not* labeled by a taxon in s, we have the single-variable clause $\neg CT_s(j)$, which ensures that leaf j will *not* be in the cluster-tree for s.

Recapping. For each cluster s in F, the forward clauses select a subset of edges of H, forming a subtree of T_s, consisting of a root node $v \neq 0$, and every node that is reachable from v using edges in T_s. A leaf will be in the set of reachable nodes *only if* it is labeled by a taxon in s.

However, with the forward clauses alone, it is *not* necessarily true that every leaf labeled by a taxon in cluster s will be reached from the chosen root, v, of the cluster-tree for s. It depends on v and topology of T_s. So, we need *backward* clauses to ensure that *every* such leaf is reached.

Backward Clauses for Requirement B: For each cluster s, we need to implement: *if $CT_s(j)$ is set *true*, *then* either j is the root of the cluster-tree for s; or for *some* node $i \neq j$, $CT_s(i)$ is *true* and $X_s(i, j)$ is true. Recall leaves cannot be roots of cluster trees.

Implementing the backward clauses in CNF is more indirect than were the forward clauses. We use a variable $Z_s(i, j)$ each directed edge(i, j) in H. $Z(s, i, j)$ will be set *true* if and only if edge (i, j) *could be* the edge entering node j in the cluster-tree for s. The following Boolean relations implement this: $CT(s, i) \wedge$

$X(s,i,j) \Rightarrow Z(s,i,j)$, and $Z(s,i,j) \Rightarrow (CT(s,i) \wedge X(s,i,j))$, which can be converted to CNF, similar to what we have done earlier.

Then, we implement the backward requirement for a cluster s and node j as:

$$[Z(s,i_1,j) \vee Z(s,i_2,j) \vee ... \vee Z(s,i_k,j)] \vee RCT(s,j) \vee \neg CT(s,j).$$

Specifying $\mathcal{D}$: For each edge (i,j) in H, the CNF for the TFP will use Boolean variable $Y(i,j)$ to indicate whether edge (i,j) will be in $\mathcal{D}$. For each cluster s, we have the clause $Y(i,j) \vee \neg X(s,i,j)$, implementing the relation $X(s,i,j) \Rightarrow Y(i,j)$. In the opposite direction, for each edge (i,j), we have the clause $[X(s,i_1,j) \vee X(s,i_2,j) \vee ... \vee X(s,i_k,j)] \vee \neg Y(i,j)$, implementing the relation that $Y(i,j)$ can be set true *only if* edge (i,j) is in a leaf-spanning tree for some cluster s.

3.1 Identifying, Counting and Limiting Reticulation Nodes in $\mathcal{D}$

Now we discuss the most difficult part of the CNF formulation for the TFP. We imagine that the values of all the Y variables have been assigned, and that we will do a sequential *scan* through the nodes in H to determine which nodes in $\mathcal{D}$ are reticulation nodes, to impose the in-degree constraint of two, and to impose the constraint that there be at most t reticulation nodes.[5] During a scan of a node j, we must do a *sub-scan* of all the edges into j in H. We also need a "dummy leaf" d (with no in or out edges), which comes after the last leaf in the ordering of nodes in H.

To count the number of reticulation nodes, we use RT variables. $RT(1,0)$ is the variable that indicates that *just before* node 1 is examined (in the imagined scan), *no* reticulation nodes have been identified. It is set *true* in the CNF formulation. In general, variable $RT(j,b)$ indicates whether *at least* b reticulation nodes have been identified, just before node j is examined. Then, for each node j and each feasible value for b, we have the relation $RT(j,b) \Rightarrow RT(j+1,b)$, meaning that if at least b reticulations have been observed just before examining node j, then at least b reticulations will be observed just before examining node $j+1$.

We also use Boolean variables $I(i,j)$ and $RI(i,j)$, where (i,j) is a directed edge in H. $I(i,j)$ will be set *true* if, when doing a scan of node j, and just before examining edge (i,j), *at least* one edge into node j has already been found to be in $\mathcal{D}$. $RI(i,j)$ has a similar meaning, but *at least* two edges into j have already been found to be in $\mathcal{D}$. That is, node j has already been determined to be a reticulation node in $\mathcal{D}$. The logic for this is: $Y(i,j) \Rightarrow I(i+1,j)$, and $(Y(i,j) \wedge I(i,j)) \Rightarrow RI(i+1,j)$, which are implemented in CNF as $I(i+1,j) \vee \neg Y(i,j)$ and $RI(i+1,j) \vee \neg Y(i,j) \vee I(i,j)$, respectively. We also need $I(i,j) \Rightarrow I(i+1,j)$ and $RI(i,j) \Rightarrow RI(i+1,j)$.

---

[5] Of course, there is no sequentiality in a CNF formula, as the values of all the variables are essentially set together. But, to understand the logic, it helps to imagine a sequential scan of the nodes, after the Y variables have been set.

Next, we use variable R to record whether a node in $\mathcal{D}$ is a reticulation node. The logic is $(Y(i,j) \wedge I(i,j)) \Rightarrow R(j)$. At first, it may seem that RI variables could just be replaced by R variables, but we still need RI variables for the following logic: $RI(i,j) \Rightarrow \neg Y(i,j)$, which means that if we have already determined that there are two edges into j (in the imagined scan) just before we are about to examine edge (i,j), then edge (i,j) must *not* be put in $\mathcal{D}$, since the in-degree of any node in $\mathcal{D}$ is at most two. Without such a bound, we could remove the RI variables, and also thereby compute the *reticulation number*.

Next, we have the logic: $(RT(j,b) \wedge R(j)) \Rightarrow RT(j+1, b+1)$, meaning that if at least b reticulation nodes have been identified just before the scan of node j, and node j is determined to be a reticulation node, then at at least $b+1$ reticulation nodes will have been identified just before the scan of node $j+1$.

Finally, we have the single-variable clause $\neg RT(d, t+1)$, where d is the "dummy" leaf of H. This ensures that $\mathcal{D}$ contains at most t reticulation nodes. This finishes the "high-level" description of the CNF for an instance of the TFP for (S, F) with a target t. Some additional clauses are needed for programming reasons and to cover degenerate-cases, and in total there are about forty-five types of clauses used in our formulation.

4   Hybridization Networks: Reticulation Networks When Input Trees Are Specified

Now we discuss a related problem that was tackled earlier by SAT-solving in a paper by Ulyantsev and Melnik [13]. In that problem, the input is not a set of *clusters*, but a set of *trees*:

Hybridization Network Problem: Given a set of rooted *binary* trees, $\mathcal{T}_1, \mathcal{T}_2, ..., \mathcal{T}_z$, each with the same set of n labeled leaves, construct a DAG $\mathcal{D}$ with a single root and with each node of in-degree at most two, that displays each of the trees $\mathcal{T}_1, ..., \mathcal{T}_z$, *minimizing* the number of reticulation nodes.

The results in [13], using SAT with an additional combinatorial heuristic that we will discuss later, are reported in Table 1. Their CNF-formulation has 68 different *types* of clauses, and is limited to binary trees.

Reducing the Hybridization Problem to the Reticulation Problem: The first attempt is to take each input tree $\mathcal{T}_i$, and create the set of $n-1$ *clusters*, each defined by a distinct non-leaf node in $\mathcal{T}_i$. Then this collection of clusters can be input to the Minimum-Reticulation Problem. The resulting DAG will display all of the clusters in the input trees, *but* it is **not** guaranteed to display the actual trees $\mathcal{T}_1, ..., \mathcal{T}_k$.

An Easy Fix: In the Minimum Reticulation Problem, a SAT solution must specify a leaf-spanning tree T_s for each input cluster s, and T_s must have a node v that reaches *exactly* the leaves of T_s labeled by s. To use the SAT-formulation for the Minimum-Reticulation problem to solve the Hybridization-Network Problem, we change the CNF to enforce the following:

If two clusters s and s' come from the same input tree $\mathcal{T}$ (in the Hybridization problem), then the trees, T_s and $T_{s'}$, specified by a SAT solution (to the Reticulation problem), must be *identical*.

To implement this, for each input tree $\mathcal{T}$, we create a *single* leaf-spanning tree $T_\mathcal{T}$ for $\mathcal{T}$, rather than one for each cluster in $\mathcal{T}$. Then, in the CNF for a TFP problem, we replace every reference to a tree T_s to the tree $T_\mathcal{T}$, if cluster s came from tree $\mathcal{T}$. This approach *unifies* the SAT approach (and computer code) for the Minimum-Reticulation and the Hybridization-Network Problems, and also allows easy modification to solve related network construction problems, for example when the input consists of some known trees, and some clusters whose underlying trees are not known.

Why This Works. The above SAT-approach to the Hybridization problem differs from that in [13]. There the SAT formulation *explicitly* detailed the topology of each input tree. In our approach, we only detail the clusters in each input tree, and then construct one spanning-leaf tree for all the clusters in the same input tree. This works because of **The Fundamental Theorem of Trees** (see [5] for a proof and exposition): The set of clusters derived from a directed tree T *uniquely* specifies T.

So, when *all* of the clusters from an input tree $\mathcal{T}$ must be displayed by the *same* leaf-spanning tree $T_\mathcal{T}$ in $\mathcal{D}$, $T_\mathcal{T}$ *must be* $\mathcal{T}$.

Decomposition Heuristics. Two methods were developed in [1,3] that can sometimes *decompose* an instance of the Hybridization Problem into several smaller instances, that are solved separately. These methods were used in SAT-solving the Hybridization problem for two trees [2], and later for multiple trees [13]. The first method removes any leaf-labeled subtree that is found in *every* input tree. The same removal is a consequence of the procedure *clean* [5], which we routinely apply to all phylogenetic network data. So, it is used for both the Minimum-Reticulation Problem and the Hybridization Problem. The second method, which is only correct for the Hybridization Problem, is the following: If *all* of the input trees contain a common cluster s (i.e., each input tree has some node which reaches all and only the nodes in s), even if the *subtrees* that reach s are *not* identical (they have different topologies), each such subtree can be removed from its input tree. The Hybridization Problem can then be solved on the remaining parts of the input trees, and combined with a solution to the Hybridization problem with input consisting of the removed subtrees.

Since the running time of the SAT-solving approach increases more than linearly with increasing size of a CNF formulation, this decomposition (when it is possible) can be very effective in speeding up the computation.

5   Some Empirical Results

We extensively tested our SAT approach to both the Minimum Reticulation Problem and to the Hybridization Problem. The solution to the Hybridization

Problem was implemented both with and without the second decomposition heuristic. We used real Grass data, detailed below, and random tree and cluster data produced by the package MS [7], using parameters that are believed to be typical of human data. We selected trees from those produced by MS, both randomly, and by taking related trees, which better aligns with real biological data. Due to space limitations, we will only detail the results of the computations for the Grass data (see Table 1, and the explanation that follows). Results for the random data can be found on GitHub (see below).

Table 1. Results of computations with the same grass (*Poaceae*) dataset studied in [13]. The data with only two trees was also studied in [2].

Dataset	Lv	RET	HYB	HYB-RET	HYB-RET-H
3NdhfPhytRpoc	22	6, 1384 (384)	8, T	8, 1227 (169)	8, 522 (3.54)
3PhytRbclRpoc	16	5, 1076 (16.29)	6, 11	6, 6644 (206)	6, 22.9
3RbclWaxyIts	12	5, 1008 (6.77)	6, T	6, 1835 (53.6)	6, 523 (14.6)
*4NdhfRbclWaxyIts	12	5, 209 (6.4)	7, T	6, 1274 (219)	6, 1631 (343)
**4PhytRbclRpocIts	15	7, 331.4 (24.6)	9, T, [8]	8, 2319 (478.6)	8, 877 (365.8)
2RbclRpos	27	7, (UB)	7, T	7, 2372 (536.7)	7, 260 (0.52)
3NdhfWaxyIts	16	7, 4313 (122.8)	8, T	8, 3588 (422.5)	8, 532 (15.8)
**3PhytRbclIts	18	7, 1173 (170.6)	11, T, [8]	8, 6470 (403)	8, 632 (124)
3PhytRpocIts	20	7, 1693 (689)	7, T	7, 6490 (430)	7, 370 (104.5)
*4NdhfPhytRbclRpocIts	16	6, 1765 (760.7)	10, T	9, 7072 (942)	9, 573 (65.7)
*4NdhfPhytRpocIts	20	8, 1099 (87.7)	10, T	9, 8805 (2573)	9, 894 (368)
2NdhfPhyt	41	8, 3000, (UB)	8, 12	12, (37595)	8, 9.2
2NdhfRbcl	37	8, 1017 (UB)	8, 1	8, 15084 (8947)	8, 1
2PhytIts	31	8, 1556 (UB)	8, 41	NS, 2000	8, 7.66
3NdhfPhytRbcl	22	7, 2494 (452)	9, 123	9, 7516 (4331)	9, 47.6
2NdhfRpoc	35	9, 1054 (UB)	9, 954	9, (8163)	9, 225.3
*3NdhfRbclRpoc	27	11, 1247 (225.7)	13, T	11, 23243 (2255)	11, 494 (224)
3NdhfPhytIts	31	12, 2323 (82)	13, T	13, 14380	13, 1773 (524)

Explaining Table 1 The number at the start of the dataset name is the number of trees in the dataset, and the next column, labeled *Lv*, gives the number of leaves in those trees. The column labeled *RET* shows our results for the *Minimum Reticulation* problem. Using the SAT solver *CryptominiSAT5* and four cores on an Apple imac, all computations reached a time-out (different for different test-cases). The first number is the number of reticulations used in the best network constructed, and the next is the total time used before time-out. When there is a third number, in parenthesis, it is the time used to find the network with the number of reticulations reported by the first number. The differences in times show that most of the time was used in the last TFP instance, which we believe is an unsatisfiable instance that times out. Entries containing "UB", indicate that the best network was found by SHRUB or KWARG, before any attempt at SAT-solving. The next column, labeled *HYB* contains the results reported in [13]. The first number is the number of reticulations. The next entry is either a "T", indicating that the computation timed-out (1000 s seconds), or is the time (in seconds) taken when the computation ran to optimality. The two entries of [8] indicate that a solution with 8 reticulations was found by the program in [14], rather than by SAT solving. The next column, labeled *HYB-RET* reports our solutions, found by reducing the Hybridization Problem to the Reticulation Problem. The first entry is either the number of reticulations found, or "NS" indicating that no TFP solution was found before timing out. The next number is the total time taken in the computation. When there is a third number, in parentheses, this is the time used to find the network with the number of reticulation nodes reported in the first number. The next column, labeled "HYB-RET-H" reports our solutions when the decomposition heuristic is also used. With that heuristic, five of the datasets are solved to optimality, exactly the same datasets that were solved to optimality in [13]. In some cases the use of the decomposition heuristic lead to a *spectacular* reduction in computation time. For example, the case of 2NdhfRbcl, where an optimal solution was found in one second, but without the heuristic, HYB-RET used 8,947 s to find the same solution, without proving optimality.

Datasets where we reduced the number of reticulations used, compared to the best previous results, are shown in bold. A single asterisk indicates that the number of reticulations for the Hybridization Problem is smaller than the previous smallest number (shown in the HYB column). There are four such cases. A double asterisk indicates that the number of reticulations is smaller than the previously smallest number obtained with a SAT formulation, but equal to what was known from non-SAT methods in [14].

5.1   Summary of the Empirical Conclusions

1) ILP is essentially useless for the network problems studied in this paper. 2) The SAT approach is practical for finding good reticulation networks for the Grass data, quickly finding good networks with many fewer reticulations than found by ILP. 3) For both the Minimum Reticulation Problem and the Hybridization Problem, the time needed to prove that a SAT instance is *Un*satisfiable is huge in comparison to the time needed when an instance is satisfiable. 4) For the Minimum Reticulation Problem, with the Grass data, the SAT approach never proved optimality (by proving a TFP instance is unsatisfiable), but the networks found are likely optimal because a) the *satisfiable* SAT instances of the TFP were solved quickly, followed by an instance of the TFP where the SAT solver timed out - characteristic of an unsatisfiable instance; b) in all datasets, the number of reticulations is at most the number of reticulations in the solutions to the Hybridization Problem, which is required of an optimal solution to the Minimum Reticulation Problem. 5) The SAT approach is generally practical for random tree data (generated by the program MS [7], using parameters that are often proposed to emulate human data) in a wide range of data (3 trees

with up to 40 leaves; 6 trees with up to 20 leaves; and 8 trees with up to 10 leaves). 6) In the Minimum Reticulation Problem, the upper bounds provided by SHRUB and KWARG were fairly close to the solution values found by the SAT computations. 7) The Minimum Reticulation Problem is generally harder to solve than the Hybridization Problem. 8) The decomposition heuristic (when it can be applied) can be extremely effective. 9) The decomposition heuristic often does *not* apply, particularly as the number of trees and/or taxa increases, or when the trees are randomly selected. 10) The SAT approach we developed for the Hybridization Problem can handle input of *non-binary* trees, and *hybrid* problems where some input comes from trees, and some from individual clusters. 11) The results for the Hybridization problem on the grass data reduced the previous minimum number of reticulations, in several cases. 12) Comparisons of the published running times in [13] to the times for our experiments would not be meaningful due to differences in machines and solvers.

All software and datasets are available on GitHub: github.com/gusfield/SAT-solving-for-phylogenetic-networks.

References

1. Baroni, M., Semple, C., Steel, M.: Hybrids in real time. Syst. Biol. **55**, 46–56 (2006)
2. Bonet, M.L., John, K.S.: Efficiently calculating evolutionary tree measures using SAT. In: Kullmann, O. (ed.) SAT 2009. LNCS, vol. 5584, pp. 4–17. Springer, Heidelberg (2009). https://doi.org/10.1007/978-3-642-02777-2_3
3. Bordewich, M., Semple, C.: On the computational complexity of the rooted subtree prune and regraft distance. Ann. Comb. **8**, 409–423 (2005). https://doi.org/10.1007/s00026-004-0229-z
4. Brown, H., Zuo, L., Gusfield, D.: Comparing integer linear programming to SAT-solving for hard problems in computational and systems biology. In: Martín-Vide, C., Vega-Rodríguez, M.A., Wheeler, T. (eds.) AlCoB 2020. LNCS, vol. 12099, pp. 63–76. Springer, Cham (2020). https://doi.org/10.1007/978-3-030-42266-0_6
5. Gusfield, D.: ReCombinatorics: The Algorithmics of Ancestral Recombination Graphs and Explicit Phylogenetic Networks. MIT Press, Cambridge (2014)
6. Gusfield, D.: Integer Linear Programming in Computational and Systems Biology: An Entry-level Text. Cambridge University Press, Cambridge (2019)
7. Hudson, R.: Generating samples under the wright-fisher neutral model of genetic variation. Bioinformatics **18**(2), 337–338 (2002)
8. Huson, D., Rupp, R., Scornavacca, C.: Phylogenetic Networks. Cambridge University Press, Cambridge (2010)
9. Ignatieva, A., Lyngsø, R., Jenkins, P., Hein, J.: KwARG: parsimonious reconstruction of ancestral recombination graphs with recurrent mutation. Bioinformatics **37**(19), 3277–3284 (2021)
10. Matsieva, J.: Optimization techniques for phylogenetics. PhD thesis, University of California, Davis, Department of Computer Science (2019)
11. Sipser, M.: Introduction to the Theory of Computation, 3rd edn. Cengage Learning, Boston (2013)
12. Song, Y.S., Wu, Y., Gusfield, D.: Efficient computation of close lower and upper bounds on the minimum number of needed recombinations in the evolution of biological sequences. Bioinformatics, **21**, i413–i422 (2005). Bioinformatics Suppl. 1, Proceedings of ISMB 2005

13. Ulyantsev, V., Melnik, M.: Constructing parsimonious hybridization networks from multiple phylogenetic trees using a SAT-solver. In: Dediu, A.-H., Hernández-Quiroz, F., Martín-Vide, C., Rosenblueth, D.A. (eds.) AlCoB 2015. LNCS, vol. 9199, pp. 141–153. Springer, Cham (2015). https://doi.org/10.1007/978-3-319-21233-3_11

14. Wu, Y.: An algorithm for constructing parsimonious hybridization networks with multiple phylogentic trees. J. Comput. Biol. **20**, 792–804 (2013)

Feature Selection for Identification of Risk Factors Associated with Infant Mortality

André Louzada Colodette[1], Fabiano Novaes Barcellos Filho[1],
Gustavo Carreiro Pinasco[2], Sheila Cristina de Souza Cruz[3],
and Sérgio Nery Simões[4(✉)]

[1] Escola Superior de Ciências da Santa Casa de Misericórdia de Vitória,
Vitória, ES, Brazil
[2] Universidade Federal do Espírito Santo, Vitória, ES, Brazil
[3] Escola Técnica do SUS, Vitória, ES, Brazil
[4] Instituto Federal do Espírito Santo, Serra, ES, Brazil
sergio@ifes.edu.br

Abstract. In the context of infant mortality risk analyses, the application of Machine Learning techniques, like Feature Selection, can be an efficient way to increase the interpretability of data and explanation of the studied phenomenon. In this paper, we developed a Machine Learning approach to identify the main risk factors that impact the local population studied with regard to infant mortality, aiming to help professionals who deal directly with the event or with the epidemiological guidelines that may be made available from data analysis. First, we integrated the databases of the Live Birth Information System (SINASC) and the Infant Mortality Information System (SIM), between 2006 and 2019, in the city of Vitória, ES, Brazil. Then, we used feature selection methods, such as SHAP, Feature_Importance and SelectKBest, to identify the main risk factors associated with infant mortality and we compared the results obtained from applying these algorithms with the most recent results of a 2018 meta-analysis. We observed that the results achieved by the methods, especially by the SHAP method, match the results of a literature meta-analysis, in which the factors that most influenced the final outcome of mortality were Weight, APGAR, Gestational Age and Presence of Anomalies. Therefore, the use of interpretability techniques, such as SHAP, are very promising for the selection and the identification of population risk factors related to infant mortality, by using existing databases without the need for new population studies and, in addition, this knowledge can be used to help in decision making for public health.

Keywords: Infant mortality · Risk factors · Data mining · Feature selection · Explainable artificial intelligence

1 Introduction

Infant mortality is the number of deaths of children under one year of age, per thousand live births, in a given geographic space and time considered. The

Supported by FAPES (T.O. 179/2019), IFES, EMESCAM and PMV – ES, Brazil.

M. S. Bansal et al. (Eds.): ICCABS 2021, LNBI 13254, pp. 92–102, 2022.
https://doi.org/10.1007/978-3-031-17531-2_8

infant mortality rate is an important indicator of maternal and child health, capable of reflecting socioeconomic and environmental conditions and factors related to prenatal care, childbirth and newborns [1]. Its monitoring is of great value to identify changes in these scenarios, in addition to being fundamental in the orientation, development and implementation of effective strategies, aimed at further reducing this, which is still significant in our country [2].

In Brazil and worldwide, in recent decades, there has been a significant reduction in this indicator, mainly attributed to progress in social, environmental and health services conditions [3]. In numbers, there was a drop in infant mortality rate from 47/1,000 live births in 1990 to 14.6/1,000 live births in 2012 [4]. Further on, in 2018 the mortality rate in the country dropped to 8/1,000 births [5]. However, the WHO plan for 2030 is to zero the infant mortality rate.

It was shown in a study that the use of already consolidated national databases and a hierarchical model allowed the integrated analysis of socioeconomic, behavioral and health risk factors, the Apgar score and biological risk factors, which can support the development of more comprehensive and contextualized public policies. Given the hypothesis that socioeconomic, environmental and birth conditions influence the outcome of death in newborns and infants, it is described the development a model that involves prenatal, peri-partum and socioeconomic characteristics for the selection of the main attributes sandrisk factors associated with predicting fant mortality [3].

Thus, many researchers and health professionals use meta-analyses, which tend to be costly and time-consuming, to better understand which are the main variables that have the greatest impact on infant mortality in the population studied. In this work, we propose the use of machine learning models focused on interpretability, as a cheaper and faster alternative approach to meta-analysis.

2   Method

2.1   Study Design and Casuistic

We used the retrospective cohort study, which includes all cases of infant deaths from 0 (zero) to 1 (one) year of age in both sexes, extracted from the Mortality Information System (SIM) and integrated with the data from the System of Information on Live Births (SINASC) registered at the Municipal Health Department of Vitória (SEMUS) between the years 2006 to 2019. The data on infant deaths came from the intersection between the SIM and SINASC databases, prospectively, from the deterministic linkage method of the data verified by the City's Epidemiological Surveillance Management. Notifications of live births or infant deaths that did not dissipate complete data for analysis of the variables of interest were excluded.

This type of study design is excellent when the predictor variables are costly and can be measured at the end of the study in subjects who developed the outcome (cases) and in a sample of those who did not (controls). The advantage of this study design is the low cost of execution since the cohort data have already been collected and stored according to the workflow of the Municipal

Health Department of Vitória and the national regulation of data collection and information from SIM and SINASC.

Data from the Municipal Health Department were accessed after the project was approved by the Ethics and Research Committee (CEP), with opinion: 3,280,796, in institutional research and after it had been processed by the SUS Technical School (ETSUS).

2.2 Database Integration

The infant death and live birth notification databases were integrated by the deterministic linkage technique, based on the agreement between the newborns identified in both databases. As the data came from an Epidemiological Surveillance Center, the identification to link birth and death information involved even more precise personal information. After the integration of the databases of the two systems, in a single database, it underwent an analysis, cleaning and preparation of relevant rows and columns, as described below.

2.3 Data Analysis, Cleaning and Preparation

While reading the database, some null values were replaced by the appropriate null value representation (NA), based on the data dictionary provided by the systems of the source databases, as some code was used in the database to represent that a certain value was non-existent. Still based on the data dictionary, the column data types were fixed.

In the integrated database, which initially had 248 columns and 173,353 records, the columns from the Mortality Information System were excluded, as this database was used only to identify which of those born had died within one year of life, thus the database was reduced by 88 columns, now having 160 columns. Afterwards, another 52 columns from the Live Births Information System were excluded, leaving 108 attributes. Finally, 63 columns that do not have medical or epidemiological importance, such as: name, record number, according to the data dictionary, were excluded, leaving 45 columns in the database.

Columns that had a large number of null values (NA) were also excluded. As a criterion, all 18 columns that had more than 20% of null values were excluded, when considering the death class, leaving 27 columns. A further 11 columns had to be excluded, as they presented values that were duplicated with other columns, or that were irrelevant, with data not related to health, such as codes, or that in pre-analysis and pre-processing proved to be irrelevant. Then, all records (rows) that had some null values were excluded, leaving the database with 155,058 records, of which 154,418 (99.59%) are from the living class, and 640 (0.41%) from the death class.

The 14 columns analyzed were: presence of anomaly, APGAR1, APGAR5, number of prenatal consultations, mother's marital status, mother's age, type of delivery, weight, number of living and dead children, weeks of gestation, sex, type of pregnancy and outcome of death. Details are exemplified in Table 1. Two additional auxiliary columns, date of birth and age of death, were also kept. The records were randomly divided, but keeping the proportion, in three new data frames: training with 93,034 (60%) records, testing with 31,012 (20%) and validation with 31,012 (20%) records.

2.4 Identification of Main Variables

To identify which variables most impacted the model, and to bring greater reliability and better explain its importance and predictive potential, we can apply methods such as SHAP (*SHapley Additive exPlanations*), feature_importances (from Scikit-Learn tree models) and selectKBest (also from Scikit-learn, based on statistical methods such as chi2). All tools mentioned above are valid for analyzing the most important features for prediction. However, they were used so that we could combine predictive power and explainability of the models.

In this context, the selectKBest method was the first used to select the best characteristics, with its statistical method, which can be evaluated before training the model. In the same vein, when training the first model, the feature_importances of the Scikit-Learn tree algorithms was used, in which the training of the reference model (more basic, before hyper-parameterization) was considered and compared with the techniques of selectKBest and later of SHAP, the latter used with the best model after training and validating it in data never seen before.

The SHAP technique uses game theory approach to explain the output of machine learning models [6]. Through a kind of reverse engineering, it connects and groups the variables and calculates a score proportional to that factor's contribution to achieving the goal, based on Shapley's classic game theory values and their related extensions [7]. The main machine learning (ML) model used as the basis for SHAP was XGBoost (Extreme Gradient Boosting) [8], due to the fact that it is the model with the best results, and it is the most indicated and used today, when it comes to structured or tabulated data.

XGBoost is a supervised model of machine learning algorithm, based on a decision tree and using a Gradient boosting structure. It builds several decision trees (which are basically like decision flowcharts). Each of the trees, individually, has a low predictive power but, with the power of the gradient, they adjust the errors through the previous residuals and pass the results to the next trees increasing their efficiency. So that the final result is an average of the results from the set of trees.

In order to increase the predictive capacity of the model, it went through a "tuning" to determine which configuration parameters best fit the data used and the project objective. The final selection of the variables of interest was confronted with the main characteristics found in the last systematic review and meta-analysis on the subject [9].

2.5   Tools, Techniques and Software Used

To carry out the research, the Python programming language in version 3.7.10 of the Jupyter development environment was used. Among the main libraries used are Pandas, for storing and manipulating tabular data, Numpy for matrix operations, and Scipy with statistical and scientific functions. Machine learning models were imported from Scikit-Learn [10], including: Decision Tree Classifier, Logistic Regression, KNeighbors Classifier, SVC, Random Forest Classifier, XGB Classifier, MLP Classifier, Gaussian NB, Gradient Boosting Classifier, Ada Boost Classifier, SGD Classifier [10]. To identify the variables, the following tools were used: SelectKBest (Scikit-Learn), feature_importances_ (Scikit-Learn), SHAP.

3   Results

3.1   Data Description

Initially, the database had 173,353 records of live births, between the years 2006 to 2019, but after cleaning and preparation, it now has 155,058 records, of which 640 (0.41%) died within one year of life. As we can see, the most prevalent cases of death were, at birth, 413 (64.53%) with low weight (less than 2500g), 89 (13.91%) who had APGAR5 less than seven, 391 (61.09%) who were born at preterm (less than 37 weeks of gestation), and 327 (50.47%) who had less than 6 prenatal consultations. Out of those who died, 404 (63.12%) were in the neonatal period (up to 28 days). The summary of the record data description can be viewed in Table 1.

3.2   Feature Selection

The interpretability of the model was performed using three methodologies and comparing the identification of risk factors. The SelectKBest, Feature Importance and SHAP (SHapley Additive exPlanations) methods identified several similar risk factors for early death, as shown in Table 2, which reinforces the importance of these characteristics for predicting infant death.

Table 1. Description of data after cleaning and preparation.

Feature		Total	Live	Death
Records		155058	154418 (99.59%)	640 (0.41%)
Anomaly	Present	871 (0.56%)	778 (0.5%)	93 (14.53%)
	Absent	154187 (99.44%)	153640 (99.5%)	547 (85.47%)
APGAR1	Average (DP)	8.14 (±0.99)	8.15 (±0.97)	6.07 (±2.29)
	≥7	147383 (95.05%)	147036 (95.22%)	347 (54.22%)
	<7	7675 (4.95%)	7382 (4.78%)	293 (45.78%)
APGAR5	Average (DP)	9.05 (±0.71)	9.05 (±0.7)	7.8 (±1.69)
	≥7	154174 (99.43%)	153623 (99.49%)	551 (86.09%)
	<7	884 (0.57%)	795 (0.51%)	89 (13.91%)
Prenatal consultation number	None	1160 (0.75%)	1135 (0.74%)	25 (3.91%)
	1 to 3	4550 (2.93%)	4460 (2.89%)	90 (14.06%)
	4 to 6	29422 (18.97%)	29214 (18.92%)	208 (32.5%)
	7 or more	119926 (77.34%)	119609 (77.46%)	317 (49.53%)
Mother's marital status	Single	73612 (47.47%)	73251 (47.44%)	361 (56.41%)
	Married	74167 (47.83%)	73918 (47.87%)	249 (38.91%)
	widow	374 (0.24%)	372 (0.24%)	2 (0.31%)
	Separated/ Divorced	3162 (2.04%)	3146 (2.04%)	16 (2.5%)
	Stable union	3743 (2.41%)	3731 (2.42%)	12 (1.88%)
Mother's age (years)	Average (DP)	27.55 (±6.5)	27.55 (±6.5)	27.52 (±7.04)
	>35 years	18659 (12.03%)	18571 (12.03%)	88 (13.75%)
	25–35 years	83301 (53.72%)	82975 (53.73%)	326 (50.94%)
	<25 years	53098 (34.24%)	52872 (34.24%)	226 (35.31%)
Childbirth	Vaginal	46396 (29.92%)	46159 (29.89%)	237 (37.03%)
	Cesarean	108662 (70.08%)	108259 (70.11%)	403 (62.97%)
Weight (g)	Average (DP)	3216.01 (±547.73)	3221.24 (±538.23)	1955.69 (±1093.38)
	Normal (2500 a 4000 g g)	134486 (86.73%)	134064 (86.82%)	220 (34.38%)
	Low Weight (<2500 g)	12319 (7.94%)	11906 (7.71%)	413 (64.53%)
	Macrosomic (>4000 g)	8253 (5.32%)	8246 (5.34%)	7 (1.09%)
Gestation age	Term (37–42 week)	140579 (90.66%)	140332 (90.88%)	247 (38.59%)
	Preterm (<37 week)	12810 (8.26%)	12419 (8.04%)	391 (61.09%)
	Post-Term (>42 week)	1669 (1.08%)	1667 (1.08%)	2 (0.31%)
Sex	Male	79497 (51.27%)	79148 (51.26%)	349 (54.53%)
	Female	75561 (48.73%)	75270 (48.74%)	291 (45.47%)
Type of pregnancy	Only	151461 (97.68%)	150895 (97.72%)	566 (88.44%)
	Twins	3499 (2.26%)	3428 (2.22%)	71 (11.09%)
	Triplets or more	98 (0.06%)	95 (0.06%)	3 (0.47%)
Occurrence of death	Neonatal period (up to 28 days of life)			404 (63.12%)
	Post-neonatal period (29 days to 365 days of life)			236 (36.88%)

Table 2. Order of risk factor of different techniques in comparison with literature.

Ranking	Literature results		Select Kbest	Feature importance (random forest classifier)	Shap (XGB classifier)
	Meta-analysis [9]	Sp model shap [11]			
1	Weight (<1500 g)	APGAR5	Presence of anomalies	Weight	Weight
2	Presence of Anomalies	Weight	APGAR1	Maternal age	APGAR1
3	APGAR5 (<7)	APGAR1	APGAR5	APGAR1	APGAR5
4	Gestational age (<37 weeks)	Presence of anomalies	Number of Pre-Christmas Consultations	APGAR5	Gestational age
5	Intercurrences during pregnancy	Gestational age	Mother's marital status	Number of living children	Presence of anomalies
6	Prenatal (absence)	–	Type of childbirth	Gestational age	Mother's marital status
7	Multiple pregnancy	–	Weight	Number of Pre-christmas consultations	Number of living children
8	Mother Without a Partner	–	Number of dead children	Mother's marital status	Number of pre-christmas consultations
9	Presence of previous stillbirth	–	Number of living children	Sex	Type of childbirth
10	Education	–	Gestational age	Type of childbirth	Maternal age

The epidemiological variables with greater strength in predicting neonatal and infant death up to 1 (one) year found were: birth weight, APGAR5, number of gestational weeks, presence of congenital anomalies and number of prenatal consultations, which is corroborated by the results obtained by other studies [4,9]. The main variables found in our results are also corroborated by the findings of other researchers, who found them with greater relevance the following variables: low birth weight and preterm birth [2,4]. Some factors significantly associated with mortality were also found: age of mother over 35 years old, male, multiple pregnancy, absence of a partner, complications of pregnancy and Cesario's delivery [9].

It was observed that the highest concentration of deaths occurred during the first 28 days of life, with a total of 484 cases, corresponding to 65% of registered deaths. Neonatal deaths in the first six days are mainly caused by maternal factors, complications of pregnancy and childbirth [1]. Researchers conducted a study of the first 28 days of life of the population of São Paulo and found the following variables as the most important for the prediction: APGAR5, birth weight, APGAR1, presence of congenital anomaly and gestational age – which are also results that corroborate our findings [11] (Fig. 1).

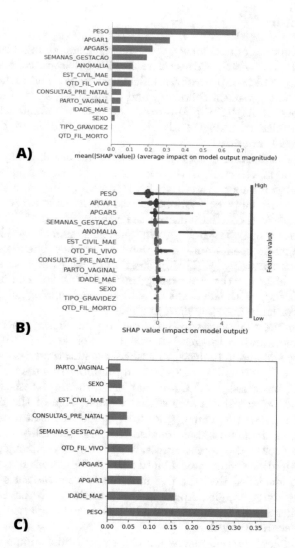

Fig. 1. Results generated by the algorithm in the selection of the main variables. A) SHAP: Average Impact of each variable on XGBClassifer. B) SHAP: Density dispersion of SHAP values for each resource on XGBClassifer. C) Feature Importance Result by RandomForestClassifier Model. **Caption:** *PESO* (Weight at Birth), *APGAR1* (APGAR in the First Minute of Life), *APGAR5* (APGAR in the Fifth Minute of Life), *SEMANAS_GESTACAO* (Gestational Age) *ANOMALIA* (Presence of Congenital Anomaly), *EST_CIVIL_MAE* (Mother's Marital Status), *QTD_FIL_VIVO* (Number of Living Children), *CONSULTAS_PRE_NATAL* (Pre-Christmas Consultation Number), *PARTO_VAGINAL* (if the delivery was via Vaginal or Cesarean), *IDADE_MAE* (Mother's Age), *SEXO* (Sex) *TIPO_GRAVIDEZ* (type of pregnancy). *QTD_FIL_MORTO* (Number of Dead Children),

4   Discussion

In the 21st century era of technology, with the expansion of big data and artificial intelligence (AI), the healthcare sector realized the need to collect and organize its data digitally, which improves task efficiency, enables and facilitates research and decision-making, patient follow-up and monitoring, and, of course, the use and development of AI [12,13]. However, the lack of infrastructure to collect the variables needed to train the algorithms makes it difficult to use machine learning in clinical practice.

Meta-analysis studies are the top of the pyramid of reliability in the results, but to allow a good meta-analysis there is a need for broad results and usually the selection of a large number of studies with technical quality and standardization in the method, which can cost a lot of time and resources to carry out the original studies and to select the articles during the review process [14].

The application of Data Mining techniques such as feature selection, prediction of deterministic events and to explain the studied phenomenon can be a more efficient and faster way. The application of algorithms to select the best characteristics related to the event using the SHAP method can help professionals who deal directly with the event or with the epidemiological guidelines that can be made available from the data analysis.

In line with the characteristics identified by the SHAP method, which can be seen in Table 1 and in Images 1-A and 1-B, Veloso et. (2019), in their systematic review and meta-analysis, identified consultations (absent), low birth weight, presence of anomalies, APGAR5 and number of weeks of gestation, the ones that bring more risk, and those that bring medium risk of death are the mother's education, marital status of the mother, Age of the mother, gender of the newborn (NB), pregnancy (multiple, single), consultations (inadequate prenatal care), as the factors most related to infant mortality [9].

Baptist et. (2021) was another researcher who used machine learning and the SHAP method to predict and identify the main factors related to neonatal mortality, who identified that the five most important variables were Apgar in the 5th minute, birth weight, Apgar in the 1st minute, presence of congenital anomaly and gestational age, respectively [11]. The comparison of the results found can be seen in Table 2.

The methodology presented was applied to birth and infant mortality data, but it can be used in different areas and health problems. For example, to find risk factors related to COVID-19 (Sars-CoV-2) [15]. As a recent disease with unpredictable prognosis, it was necessary to find patterns among patients, in which traditional studies would be infeasible due to the delay in their performance.

Another application of these techniques is the individualization of the risk of cardiac patients, which has already surpassed in accuracy and interpretability the score created by the large study by Framingham [16].

Therefore, the use of interpretability techniques for Machine Learning models, such as SHAP, can be a good way to select and identify population characteristics related to infant mortality, using existing databases and without the need for new population studies. The use of these algorithms in loco can be of great help in decision making at the public health level.

For data integration, deterministic linkage with personal variables has greater strength. However, when there is a lack of more precise information, it is possible to forego the use of probabilistic linkage, which has been widely used to monitor outcomes in cohort studies. This allows you to integrate databases of a different nature, even in the absence of a unique identifier field. This is possible through the joint use of common fields in related bases to estimate the probability that a pair of records refers to the same individual.

5   Final Considerations

It is also important to emphasize that the variables identified as risk factors by the techniques described in this study for prediction do not apply as a causal relationship. Since, in an observational study, correlated variables do not imply causality, due to the existence of hidden variables, in addition to the randomness factor that also interfere with the outcome. In other words, the outcome cannot be attributed only to the variables identified in the study. Finally, the continuation of this study is the development of an Artificial Intelligence (AI) algorithm to predict the births with the highest probability of dying within 1 year of life, and thus alert the physician and health professionals.

Funding. The authors would like to thank the FAPES (*Fundação de Amparo à Pesquisa do Espírito Santo*) for its sponsorship. We also thank the PMV-ES (*Prefeitura Municipal de Vitória do Espírito Santo*) for granting us access to their data.

References

1. Kassar, S.B., Melo, A.M., Coutinho, S.B., Lima, M.C., Lira, P.I.: Determinants of neonatal death with emphasis on health care during pregnancy, childbirth and reproductive history. J Pediatr. (Rio J) **89**(3), 269–77 (2013). https://doi.org/10.1016/j.jped.2012.11.005. PMID: 23680300
2. Borgesa, T.S., Vayego, S.A.: Risk factors for neonatal mortality in a county in Southern region. Ciência Saúde (Paraná) **8**(1), pp. 7–14 (2015). https://doi.org/10.15448/1983-652X.2015.1.21010
3. Garcia, L.P., Fernandes, C.M., Traebert, J.: Risk factors for neonatal death in the capital city with the lowest infant mortality rate in Brazil. J. Pediatr. (Rio J) **95**(2), 194–200 (2019). https://doi.org/10.1016/j.jped.2017.12.007
4. Gaiva, M.A.M., Fujimori, E., Sato, A.P.S.: Maternal and child risk factors associated with neonatal mortality. Texto Contexto Enferm **25**(4), e2290015 (2016). https://doi.org/10.1590/0104-07072016002290015
5. World health statistics 2020: monitoring health for the SDGs, sustainable development goals. Geneva: World Health Organization (2020). Licence: CC BY-NC-SA 3.0 IGO
6. Welcome to the SHAP documentation [Internet]. Welcome to the SHAP documentation - SHAP latest documentation. https://shap.readthedocs.io/en/latest/

7. Lundberg, S.M., Lee, S.-I.: A unified approach to interpreting model predictions. In: Proceedings of the 31st International Conference on Neural Information Processing Systems (NIPS 2017), pp. 4768–4777. Curran Associates Inc., Red Hook (2017)
8. XGBoost Documentation [Internet]. XGBoost Documentation - xgboost 1.5.0-dev documentation. https://xgboost.readthedocs.io/en/latest/
9. Veloso, F.C.S., Kassar, L.M.L., Oliveira, M.J.C., et al.: Analysis of neonatal mortality risk factors in Brazil: a systematic review and meta-analysis of observational studies. J. Pediatr. (Rio J) **95**(5), 519–530 (2019). https://doi.org/10.1016/j.jped.2018.12.014
10. Pedregosa, F., et al.: Scikit-learn: machine learning in Python. JMLR **12**, 2825–2830 (2011)
11. Batista, A.F.M., Diniz, C.S.G., Bonilha, E.A., Kawachi, I., Chiavegatto Filho, A.D.P.: Neonatal mortality prediction with routinely collected data: a machine learning approach. BMC Pediatr. **21**(1), 322 (2021). https://doi.org/10.1186/s12887-021-02788-9
12. Panch, T., Mattie, H., Celi, L.A.: The "inconvenient truth" about AI in healthcare. NPJ Digit Med. **2**, 77 (2019). https://doi.org/10.1038/s41746-019-0155-4
13. Hamet, P., Tremblay, J.: Artificial intelligence in medicine. Metabolism **69S**, S36–S40 (2017). https://doi.org/10.1016/j.metabol.2017.01.011
14. Hernandez, A.V., Marti, K.M., Roman, Y.M.: Meta-analysis. Chest **158**(1S), S97–S102 (2020). https://doi.org/10.1016/j.chest.2020.03.003
15. Fernandes, F.T., de Oliveira, T.A., Teixeira, C.E., et al.: A multipurpose machine learning approach to predict COVID-19 negative prognosis in São Paulo, Brazil. Sci. Rep. **11**, 3343 (2021). https://doi.org/10.1038/s41598-021-82885-y
16. Alaa, A.M., Bolton, T., Di Angelantonio, E., Rudd, J.H.F., van der Schaar, M.: Cardiovascular disease risk prediction using automated machine learning: a prospective study of 423,604 UK Biobank participants. PLoS One **14**(5), e0213653 (2019). https://doi.org/10.1371/journal.pone.0213653

Addressing Classification on Highly Imbalanced Clinical Datasets

Alexandre Babilone Fonseca[1], David Correa Martins-Jr[2], Zofia Wicik[2,3], Marek Postula[3], and Sérgio Nery Simões[1(✉)]

[1] Federal Institute of Espírito Santo (IFES), Serra, ES, Brazil
sergio@ifes.edu.br
[2] Federal University of ABC (UFABC), Santo André, Brazil
david.martins@ufabc.edu.br
[3] Department of Experimental and Clinical Pharmacology, Center for Preclinical Research and Technology CEPT, Medical University of Warsaw, Warsaw, Poland
{zofia.wicik,mpostula}@wum.edu.pl

Abstract. During the last twenty years, machine learning provided a myriad of frameworks and tools to improve data analyses in several fields. Classification, regression, clustering and dimensionality reduction techniques have been widely used in clinical studies to assist health professionals in screening, risk estimation, diagnostics and prognostics. Prospective studies often involve a long follow-up period and a large sample, therefore many investigations rely on a retrospective technique to develop precise classifiers. However, biological data usually presents a limited number of samples and imbalanced number of classes, which affects classification performance. These issues can be alleviated by employing balancing techniques, which increase the number of samples of the minority classes (oversampling) and/or decrease the number of samples of the majority classes (undersampling). In this work, we propose an original framework to assess several balancing techniques, combining them with 3 out-of-the-box classifiers. We applied the combination of techniques to the AVOCADO clinical study, which consists of a database of patient information including cardiovascular death or survival. Our results from the retrospective analysis of this database showed that for training the algorithm to predict cardiovascular outcomes in both sexes, the best undersampling techniques were ENN, RENN and Near-Miss 3, while ADASYN and SMOTE were the best oversampling techniques. Regarding the classifier algorithms, Random Forest and Logistic Regression (with internal balancing parameter enabled) achieved the best results with both families of balancing techniques. Proper balancing techniques associated with feature importance analysis improved the identification of clinical patterns in the data, enabling detection of high risk patients. This approach can be used for personalized medicine, for improving patients survival and recovery.

Supported by organizations IFES, CAPES and FAPESP (procs 2018/18560-6, 2018/21934-5) and EMPATHY trial ABM05/2020/1.1.

M. S. Bansal et al. (Eds.): ICCABS 2021, LNBI 13254, pp. 103–114, 2022.
https://doi.org/10.1007/978-3-031-17531-2_9

Keywords: Clinical datasets · Highly imbalanced data · Balance techniques · Supervised classification

1   Introduction

Data analysis based on machine learning methods became ubiquitous in almost all research areas. Due to a great increase of biology and healthcare research especially in the last two decades, machine learning techniques allowed efficient knowledge extraction from the vast amount of biological and clinical data. [1, 9]. Techniques such as classification, regression, clustering and dimensionality reduction have been widely employed in clinical research to assist healthcare professionals in screening, risk estimation, diagnoses, prognoses [11]. In addition, to achieve more accurate results, researchers have discovered that the use of data integration and/or data fusion [1] has been very promising to allow a more systemic approach fostering substantial advances in the health sciences. Such a computational solutions, including Machine learning and Artificial Intelligence, can help in the implementation of the drug-patient profile matching schemes-are likely to become more prevalent in the future, and could be greatly facilitated by development of "Clinical Decision Support" tools providing insight into optimal ways of treating patients [15]. However, one of the biggest challenges of working with clinical data is handling missing and unbalanced data [5, 8]. The unbalanced data problem is even worsened by the fact that the number of clinical samples available is usually very limited [4].

Most of the retrospective clinical data present imbalance issues where the number of control samples is much larger than the number of case samples, which are generally obtained from patients with a given disease or any undesirable clinical output of interest [5, 8]. These imbalance problems made difficult to develop a classifier which achieves good prediction metrics, especially considering the hit rate of correct classification of the case samples (minority class). Data imbalance is especially a problem if model is misspecified, and there is interest in good performance on a minority class or or the model itself. If the model does not describe reality correctly, it will minimize the deviation from the most frequently observed type of samples. Thus, achieving a good overall accuracy does not mean that the algorithm performs well. In fact, for an adequate assessment of the algorithm performance, it is required an analysis of several evaluation metrics combined such as precision, recall, F1-score, F2-score, among others [5]. Especially F1 and F2 scores are more suitable in cases where successful retrieval of the minority class is of vital importance. As such we understand dealing with clinical data where the disease or anomalous output cases are usually rare and cost-sensitive, i.e., missing anomalous output cases might impose a severe risk to the patients health and dramatically increase their further treatment costs [5].

The existing data balancing techniques are applied in basically three ways: (I) Undersampling: reducing the number of samples from the majority class; (II) Oversampling: increasing the number of samples from the minority and; (III) Undersampling+Oversampling: applying undersampling followed by the oversampling techniques, or vice-versa. In addition, there are several algorithms that can be used for both (undersampling and oversampling) of the above forms, and each of them can impact differently on the results of each metric used [5,6]. Therefore, it is usually a challenge for healthcare researchers to choose the balancing technique that may be most applicable to their case studies. For this reason, further studies are needed to assess the impact of using data balancing techniques on highly imbalanced biological data. It can also assess how much each balance technique can result in an improvement on different evaluation metrics.

Since each data set requires different balancing techniques, in this work we approached the problem by proposing a framework to deal with imbalanced data, which compares a combination of popularly used classifiers such as Logistic Regression [17], Random Forest [3] and XGBoost[1], along with undersampling and oversampling techniques [5]. The results showed that the best balancing techniques could depend on the type of data, indicating that our proposed framework could be employed as a recommendation system. We adopted as case study the retrospective dataset of the AVOCADO clinical study [12], evaluating the identification of patients with increased risk of death.

2   Related Works

Imbalanced data sets analysis usually requires the application of sampling methods to change the proportion of classes to a more balanced distribution [6]. Literature shows that the classification tends to achieve better results when applied to balanced datasets compared to imbalanced data sets [2,6,16], with few exceptions though [6,7]. Regarding clinical datasets, they usually present limited sample size and unbalanced class distributions, in which most of the samples come from only one class. In particular, clinical or disease studies, for instance, samples from patients with a given disease (case) are generally much less numerous than healthy (control) samples [10,14]. In addition, clinical studies are cost-sensitive, meaning that it is imperative to successfully identify patients with a given disease or some potential to further develop a disease of interest, although the clinical datasets tend to present much more control samples than disease samples [5]. This recurrent problem motivates studies that compare and recommend combinations of classifiers, balancing techniques and parameter configurations, depending on the intrinsic characteristics of the data at hand. With this aim in mind, Van-Hulse et al. [16] provided a comprehensive study involving 11 machine learning techniques, 7 data balancing techniques and 35 datasets, showing the importance of the application of data balancing techniques to improve the classification in terms of area under the curve (AUC). Nevertheless, to the best of our knowledge, the work presented here is the first one to propose a framework to allow a systematic comparison of a myriad of combinations of classifiers, balancing techniques and parameterizations for a given input data, further recommending the top ranked combinations based on F1 and/or F2-scores.

---

[1] https://github.com/dmlc/xgboost.

3   Materials and Methods

3.1   Method Overview

The aim of this study was to identify patients which are in increased risk of earlier death. The data we used as case (is called AVOCADO, see Sect. 3.2 for details) was a retrospective clinical data. This dataset had imbalance issues related to very low number of primary outcome (death) cases, making it challenging to develop the classifier. In addition to imbalance problem, the chosen biological dataset had relatively low number of samples, which was a challenge to train the classifier.

Figure 1 shows the method overview. First, we loaded the clinical data set, then we divided the data in the sections related to the type of the data. Our workflow was applied to each section independently. For instance, we loaded the AVOCADO clinical study dataset (see Sect. 3.2 for details), and sectioned following data: demographics, treatments, laboratory results, platelet functions tests, serum biomarkers, and outcomes. In order to achieve a good statistical power, we ran the experiment by 30 rounds for each combination of classifier[2], balance techniques and parameters. For each round, we performed the holdout splitting 70% of dataset to train and 30% to test. Next, only on the train data, we applied the combined balance techniques in two ways: (1) using first oversampling and then (2) undersampling techniques.

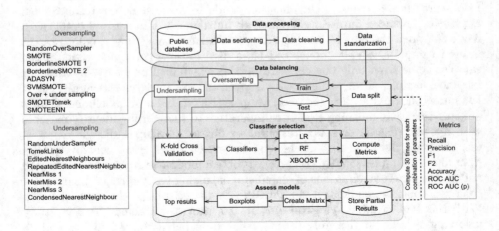

Fig. 1. Method overview

The following undersampling techniques were employed: NearMiss (versions 1, 2 and 3), Cluster Centroids, Edited Nearest Neighbours and Repeated Edited

---

[2] In this work, we used the Scikit-Learn Python library (https://scikit-learn.org/) to perform the analyses: *Scikit-learn: Machine Learning in Python*, Pedregosa et al., JMLR 12, pp. 2825–2830, 2011.

Nearest Neighbours. The oversampling techniques used were SMOTE, Borderli-neSMOTE and ADASYN. Both balancing techniques (oversampling, undersampling) have parameters which can be used for controlling the resampling ratio data set. These parameters can be defined to specify the majority reduction and minority augmentation classes intended after the balancing method application.

For the majority class, we performed the reduction on the original data to 85%, 70% and 55%. For the minority class, we increased the original number of samples to 300%, 400% and 500%. This represents a Cartesian's product of $3 \times 3 = 9$ parameter combinations. For each combination of reduction and augmentation parameters[3], the proportion of samples between the minority class and the majority class did not exceed the ratio of 1. Since some undersampling techniques like *Edited Nearest Neighbours* (ENN) and *Repeated Edited Nearest Neighbours* (RENN) do not have a way to fully control the final number of samples, in such cases, the desired ratio was not reached.

Once the dataset was balanced according to the desired ratio – for both over and undersampling , we performed *K-Fold Cross-Validation* (with K = 3) on the training set using the classifiers: *Logistic Regression, Random Forest* and *XGBoost*. After the validation of each above mentioned classifiers, we applied the validated model to the test set and computed and stored the performance metrics. Afterwards, the data splitting (which creates new training and test partitions), all subsequent processes were repeated 30 times. Finally, we computed the means of all metrics and sort by median of F2-score metric values.

3.2   Avocado Dataset and Data Cleaning

The database used in this study was obtained from the retrospective multi-center, prospective, randomized, and open-label Aspirin Versus/Or Clopidogrel in Aspirin-resistant Diabetics inflammation Outcomes (AVOCADO) study, provided by Prof. MD Marek Postula from the Medical University of Warsaw [12]. The study subjects were recruited consecutively from patients with Type 2 diabetes presenting to the outpatient clinic of the Central Teaching Hospital of the Medical University of Warsaw. The full characterization of the study population as well as the inclusion and exclusion criteria were previously published: [12,13]. The original aim of the AVOCADO study was to compare the effects of two anti-platelet treatment strategies (150 mg ASA vs 75 mg clopidogrel) on the plasma level of inflammatory markers in patients with Type 2 Diabetes Mellitus (DMT2) with high platelet reactivity (HRP). For training the algorithm we we used all cases of cardiovascular death of both sexes (primary outcome). The entire database totals 304 records (rows) and 256 attributes/features (columns).

---

[3] Regarding the parameters configuration of over-sampling techniques, to SMOTE and BorderlineSMOTE we set $k\_neighbors = 5$ and to ADASYN we set n_neighbors = 5. Concerning the parameters configuration of under-sampling techniques, to NearMiss versions 1, 2 and 3 we set n_neighbors = 3 and $n\_neighbors\_ver3 = 20$ (only for version 3). For ENN and RENN we set n_neighbors = 3 and, to ClusterCentroids, voting = 'auto'.

Since the data are diverse, the database was divided into several sections: demographics, follow-up mortality, genetic analysis, laboratory results, miRNAs analysis, platelet function tests, serum biomarkers, treatments and outcomes.

In this study, we used only the 7 sections described below:

- **Demographics:** contained demographic data of patients, such as: gender, age, height, weight, body mass index, hip measurement, systolic blood pressure, among others. Among the 21 attributes, 11 were categorical and 10 were numeric.
- **Treatments:** contained information about the use of treatments administered to patients, such as insulin, diuretics, indapamide, among others. Among the 25 attributes, 24 were categorical and 1 was numeric.
- **Laboratory results:** contained results of 32 types of laboratory tests, such as measurements of white blood cell rate, hemoglobin, hematocrit, mean blood cell volume, among others. All 32 attributes were numeric.
- **Platelet function tests:** contained results related to platelet function tests. Among the 11 attributes, 8 were categorical and 3 were numeric.
- **Serum biomarkers:** referred to serum biomarker data. All 27 attributes were numeric.
- **miRNAs:** referred to differentialy expressed miRNAs identified in microarray study (Affymetrix GeneChip miRNA 4.0 Array), and further validated in the patients plasma using qPCR.
- **Outcomes:** referred to the results (output) of the study. Only the primary end point attribute was used in our study, as the target class of all sections. It indicated the death/survival of the patient.

In order to perform the data cleaning, some exclusion criteria were adopted. Attributes with considerable indices of missing values were removed (*ratio >* 40%). Records with missing values were also deleted.

Table 1 shows the data characterization of *survived, death class* and *missing records*. The first two (*survived, death class*) are presented after the data cleaning, while the *missing records* are the numbers before data cleaning. The data cleaning was applied to all the sections of AVOCADO database: **Demographics, Treatments, Laboratory Results, Platelet Function Tests, Serum Biomarkers** and **miRNA's.** In addition to having a small number of records, the proportion between classes was unbalanced, which would greatly impair prediction. First, attributes (columns) having more than 20% of missing values were excluded. After that, rows with missing records were also excluded.

3.3   Performance Metrics

The performance metrics [5] that can be used to assess the experiments as binary classifiers are: accuracy, recall, precision, F1-score and F2-score. The recall metric defines how many positive samples were correctly predicted by the model. Precision represents the rate of correct positive predictions made. The F1-score is the harmonic mean of the precision and recall metrics. The F2-score is similar to F1-score, but it gives more weight to the recall and less weight to the precision.

Table 1. Data characterization of survived (after cleaning), death (after cleaning) class and missing records (before cleaning). The data cleaning was applied to all sections of AVOCADO database. In addition to having a small number of records, the proportion between classes was unbalanced, which would greatly impaired prediction. (*) Obs: due high number of missing record on Platelet and Serum biomarkers sections, the KNN-imputer was applied.

Section	Survived	Death	Missing
Demographics	248 (88.6%)	32 (11.4%)	22 (7.2%)
Treatments	251 (89.1%)	31 (10.9%)	18 (5.9%)
Laboratory	247 (88.9%)	31 (11.1%)	24 (7.9%)
Platelet (*)	268 (88.8%)	34 (11.2%)	75 (24.7%)
Serum (*)	268 (88.8%)	34 (11.2%)	259 (85.2%)
MiRNA's	206 (88.8%)	26 (11.2%)	72 (23.7%)

4   Results

We used original framework proposed in previous section to extract, clean, pre-process, balance the data, and at the end, to assess and store the results. We have set the experiment to run through several different configurations: 8 over-sampling techniques with 3 ratio desired parameters; 5 under-sampling with 3 ratio desired parameters, which led to roughly 400 combinations. For each above mentioned combination of parameters, we ran the analysis 30 times per classifier (Random Forest, Logistic Regression and XGBoost). In total, 37,260 results were obtained; 12,420 per classifier (see Sect. 3.1 for details). Next, we generated boxplots representing F2-score distributions, and sorted them by median values. We adopted F2-score metric because it gave more weight to recover the minority class. It reduced false negatives corresponding to missing disease cases [5].

4.1   Techniques Comparison for Demographics Data Section

Figure 2 shows F2-score boxplots for the top 10 (out of 48) combinations of oversampling and undersampling techniques, plus a baseline (no resampling technique). All of them were applied to **Demographics** section of the dataset. The boxplots were sorted by F2-score medians. We notice that there is a high variability of the results caused by different data splits (30 times per combination). As expected, the baseline (no resampling technique applied) had the worst results, suggesting the need of balancing. Among the undersampling techniques, *ENN/RENN* were highlighted with 9 out of the top 10 combinations, while *Near-Miss-3* responded by one combination in the 10th place. Regarding oversampling techniques, *SMOTE*, *ADASYN* and *BorderlineSMOTE* were the best techniques. These results show that the undersampling techniques had more influence than the oversampling techniques to achieve the best results for the considered data (Demographics section). Roughly the same pattern was observed for 5 out of 6 data sections, as discussed later in Sect. 4.3.

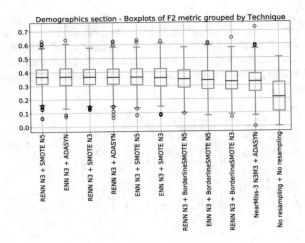

Fig. 2. F2-score boxplots for the top 10 (out of 48) combinations of balancing techniques (oversampling+undersampling) plus a baseline (no resampling technique applied). The boxplots are sorted by median of F2-metric. Each boxplot corresponds to 90 points (30 rounds × 3 classifiers).

4.2   Classifiers and Techniques Comparison for Demographics Section

We grouped 37, 260 results by classifier and technique, which led to one boxplot with 138 results per group. To simplify it, we show only the top 10 grouped results, since the main idea here is to investigate whether some combinations of classifiers and balancing techniques present more synergy. Figure 3 shows these top 10 boxplots plus the 3 baseline boxplots (no-resampling, i.e., no external balance applied) as references, one baseline for each considered classifier (LR, RF, XGB). As expected, the baselines with no balancing presented the worst results compared to the others top 10 boxplots. However, LR classifier achieved reasonable results without external balancing due to its active internal balancing mode. Regarding the top 10 boxplots, we highlight that **undersampling** *ENN/RENN* techniques were present in all of them, combined with LF or RF classifiers. Regarding **oversampling** techniques, *SMOTE, ADASYN* and *BorderlineSMOTE* were the best techniques. Thus, for the considered data (Demographics section), we can conclude that the undersampling techniques combined with LR/RF classifiers were determinant to achieve the best results. Something similar was observed for 5 out of 6 data sections (see details in Sect. 4.3).

Table 2 shows the same results presented in Fig. 3, which are the top 10 boxplots plus the 3 baseline boxplots (no-resampling), one baseline for each classifier (LR, RF, XGB) ordered by F2 medians, aggregating other metrics besides F2-score: ranking, F1-scores, precision, recall, accuracy and ROC AUC. As we can see, the rankings of LR, RF and XGB are 47, 130 and 138, respectively.

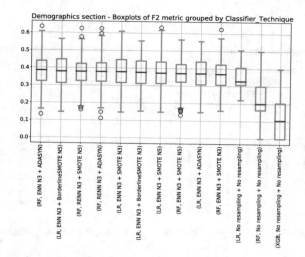

Fig. 3. Boxplots with F2-scores generated from 30 rounds of the classifiers Random Forest (RF), Logistic Regression (LR) and XGBoost (XGB), combined with balancing techniques (oversampling + undersampling), for the Demographics section. Only the top 10 results sorted by F2-score medians, plus 3 baselines (the last 3 boxplots with no resampling, one per classifier) are shown.

Table 2. All metrics results from Demographic section.

Rank	Classifier	Technique	F2	F1	Recall	Precision	Accuracy	ROC AUC
1	RF	ENN N3 + ADASYN	0.385	0.286	0.500	0.200	0.714	0.634
2	LR	ENN N3 + BorderlineSMOTE N5	0.379	0.278	0.500	0.191	0.679	0.646
3	RF	RENN N3 + SMOTE N5	0.379	0.268	0.500	0.184	0.690	0.624
4	LR	ENN N3 + SMOTE N3	0.379	0.267	0.550	0.178	0.631	0.632
5	RF	RENN N3 + ADASYN	0.379	0.265	0.500	0.182	0.673	0.618
6	LR	ENN N3 + BorderlineSMOTE N3	0.376	0.270	0.500	0.185	0.690	0.659
7	LR	ENN N3 + SMOTE N5	0.373	0.261	0.500	0.171	0.631	0.640
8	RF	ENN N3 + SMOTE N5	0.372	0.280	0.500	0.200	0.726	0.630
9	LR	ENN N3 + ADASYN	0.372	0.255	0.500	0.167	0.619	0.628
10	RF	ENN N3 + SMOTE N3	0.369	0.286	0.500	0.200	0.726	0.636
47	LR	No resampling + No resampling	0.329	0.235	0.500	0.163	0.637	0.615
130	RF	No resampling + No resampling	0.200	0.200	0.200	0.200	0.798	0.609
138	XGB	No resampling + No resampling	0.105	0.114	0.100	0.134	0.827	0.547

This means that, with no external balancing, RF and XGB are among the worst results, while LR achieved a reasonable rank due to its active internal balancing.

We also computed the Spearman correlations of all metrics pairs considering all 141 results and, as expected, *F2-score* presented a strong correlation with *recall* (0.93), a moderate correlation with *F1-score* (0.69) and a moderate negative correlation with *Accuracy* (−0.60). Therefore, balancing techniques (both external and internal) can substantially enhance the results and, when combined with a proper tuning (of classifier hyper-parameters), could further improve the results.

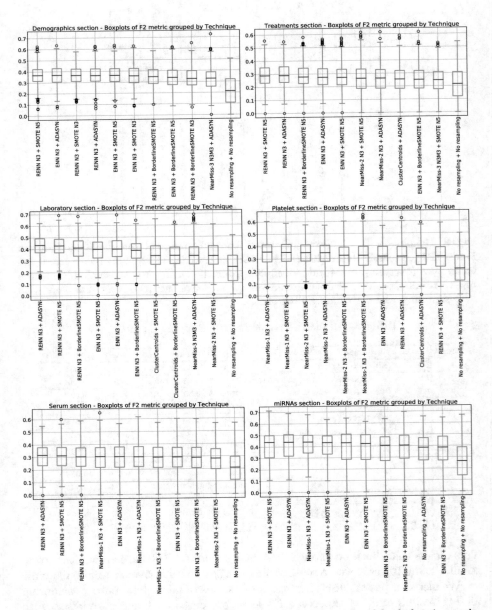

Fig. 4. Boxplots of F2-score distributions after 30 rounds grouped by balancing technique, for all six data sections. Each graph shows the F2-score boxplots for the top 10 combinations of undersampling and oversampling techniques ranked by medians, plus the baseline (without balancing).

4.3   All Data Sections Comparisons

Figure 4 shows the top 10 F2-score boxplots grouped by balancing techniques for all six sections: Demographics, Treatments, Laboratory, Platelet, Serum, and miRNA's, and one additional boxplot corresponding to the baseline (no re-sampling). As expected, the boxplot with no re-sampling techniques presented the worst results for all sections, also indicating the need for the application of balancing techniques. We also noticed that, for all graphs except one[4], there is a similar pattern on the best combination of techniques: undersampling (RENN/ENN) and oversampling (Smote/Adasyn/BorderlineSmote). The platelet section was an exception, since it presented a different pattern where the Near-Miss undersampling technique appeared in most of the top 10 results. Therefore, this shows that the best balancing techniques could depend on the type of data, which indicates that our proposed framework could be employed as a recommendation system to suggest the most appropriate balancing techniques for the considered type of dataset.

5   Conclusion

In this manuscript, we proposed an original framework to systematically investigate the classification accuracy of highly imbalanced data following distinct balancing techniques. The problem of imbalanced data is common in the biological and clinical fields, which supports the generality and relevance of our work. In order to find the best combination for dealing with the imbalanced data problem, we used the AVOCADO retrospective clinical dataset to compare the performance of oversampling and undersampling techniques with multiple distinct parameter combinations. We combined this with the three popular classifiers often used in biological area: Logistic Regression, Random Forest and XGBoosting. We found that best classification performance are obtained with undersampling techniques, especially ENN/RENN and Near-Miss-2.

The best performance were observed with out-of-the-box classifiers (i.e. without testing several different sets of parameters), and Random Forest (with internal balancing parameter enabled) and Logistic Regression classifiers. Previous studies supported that XGBoost potentially achieve excellent results; however, it also requires a lot of hyper-parameters tuning, which restricts its use by researchers without this expertise. For avoiding this restriction, we tested the classifiers with minimal hyper-parameter optimization. We notice that the best combination of classifiers and balancing techniques depends on the type of data, supporting that our framework can be employed for selecting the most appropriate techniques for each dataset.

Clinical researchers are interested in techniques which allow easy interpretation of the classification models. Our framework combined with automatized

---

[4] Platelet graph shows better results with Near-Miss techniques. This indicates the requirement for an automatic framework to find the best combination for each dataset.

interpretation analyses can improve the identification of meaningful patterns in the data. Future works should focus in the development of these tools, in order to identify high risk subjects based on critical biomarkers, improving patients survival and recovery by personalized medicine.

References

1. Ahmed, Z., Mohamed, K., Zeeshan, S., Dong, X.: Artificial intelligence with multi-functional machine learning platform development for better healthcare and precision medicine. Database **2020**, 1–35 (2020)
2. Estabrooks, A., Jo, T., Japkowicz, N.: A multiple resampling method for learning from imbalanced data sets. Comput. Intell. **20**(1), 18–36 (2004)
3. Breiman, L.: Random forests. Mach. Learn. **45**, 5–32 (2001)
4. Fehr, D., et al.: Automatic classification of prostate cancer Gleason scores from multiparametric magnetic resonance images. Proc. Natl. Acad. Sci. **112**(46), E6265–E6273 (2015)
5. Fernández, A., García, S., Galar, M., Prati, R.C., Krawczyk, B., Herrera, F.: Learning from Imbalanced Data Sets. Springer, Cham (2018). https://doi.org/10.1007/978-3-319-98074-4
6. He, H., Garcia, E.A.: Learning from imbalanced data. IEEE Trans. Knowl. Data Eng. **21**(9), 1263–1284 (2009)
7. Japkowicz, N., Stephen, S.: The class imbalance problem: a systematic study. Intell. Data Anal. **6**(5), 429–449 (2002)
8. Krawczyk, B., Galar, M., Jeleń, Ł, Herrera, F.: Evolutionary undersampling boosting for imbalanced classification of breast cancer malignancy. Appl. Soft Comput. **38**, 714–726 (2016)
9. Larrañaga, P., et al.: Machine learning in bioinformatics. Brief. Bioinform. **7**(1), 86–112 (2006)
10. Min, S., Lee, B., Yoon, S.: Deep learning in bioinformatics. Brief. Bioinform. (March) bbw068 (2016)
11. Mohedano-Munoz, M., Alique-García, S., Rubio-Sánchez, M., Raya, L., Sanchez, A.: Interactive visual clustering and classification based on dimensionality reduction mappings: a case study for analyzing patients with dermatologic conditions. Expert Syst. Appl. **171**(2019), 114605 (2021)
12. Rosiak, M., et al.: Effect of ASA dose doubling versus switching to clopidogrel on plasma inflammatory markers concentration in patients with type 2 diabetes and high platelet reactivity: the AVOCADO study. Cardiol. J. **20**(5), 545–551 (2013)
13. Sabatino, J., et al.: MicroRNAs fingerprint of bicuspid aortic valve. J. Mol. Cellular Cardiol. **134**(July), 98–106 (2019)
14. Oh, S., Lee, M.S., Zhang, B.-T.: Ensemble learning with active example selection for imbalanced biomedical data classification. IEEE/ACM Trans. Comput. Biol. Bioinform. **8**(2), 316–325 (2011)
15. Shah, P., et al.: Artificial intelligence and machine learning in clinical development: a translational perspective. NPJ Digit. Med. **2**(1), 69 (2019)
16. Van Hulse, J., Khoshgoftaar, T.M., Napolitano, A.: Experimental perspectives on learning from imbalanced data. In: Proceedings of the 24th International Conference on Machine Learning - ICML 2007, vol. 227, pp. 935–942. ACM Press, New York (2007)
17. Vapnik, V.: Statistical Learning Theory. Wiley, New York (1998)

mcPBWT: Space-Efficient Multi-column PBWT Scanning Algorithm for Composite Haplotype Matching

Pramesh Shakya[1], Ardalan Naseri[2], Degui Zhi[2], and Shaojie Zhang[1]([✉])

[1] Department of Computer Science, University of Central Florida, Orlando, FL, USA
shakyapramesh@knights.ucf.edu, shzhang@cs.ucf.edu
[2] School of Biomedical Informatics, University of Texas Health Science Center
at Houston, Houston, TX, USA
{ardalan.naseri,degui.zhi}@uth.tmc.edu

Abstract. Positional Burrows-Wheeler Transform (PBWT) is a data structure that supports efficient algorithms for finding matching segments in a panel of haplotypes. It is of interest to study the composite patterns of multiple matching segments or blocks arranged contiguously along a same haplotype as they can indicate recombination crossover events, gene-conversion tracts, or, sometimes, errors of phasing algorithms. However, current PBWT algorithms do not support search of such composite patterns efficiently. Here, we present our algorithm, mcPBWT (multi-column PBWT), that uses multiple synchronized runs of PBWT at different variant sites providing a "look-ahead" information of matches at those variant sites. Such "look-ahead" information allows us to analyze multiple contiguous matching pairs in a single pass. We present two specific cases of mcPBWT, namely *double-PBWT* and *triple-PBWT* which utilize two and three columns of PBWT respectively. *double-PBWT* finds two matching pairs' combinations representative of crossover event or phasing error while *triple-PBWT* finds three matching pairs' combinations representative of gene-conversion tract.

Keywords: PBWT · Haplotype match · Recombination · Gene conversion

1 Introduction

Durbin's PBWT (positional Burrows-Wheeler Transform) [4] is an efficient data structure that operates on a panel of haplotypes that are bi-allelic to find all the matching segments of a given minimum length threshold, L. In addition to that, it's also capable of finding matches for a query haplotype against a reference haplotype panel and other useful compression algorithms.

Given the linear time complexity of PBWT algorithms, they scale well to large datasets. Because of this, they have widely been incorporated in state-of-the-art statistical phasing and imputation tools [3,6,12]. PBWT algorithms

M. S. Bansal et al. (Eds.): ICCABS 2021, LNBI 13254, pp. 115–130, 2022.
https://doi.org/10.1007/978-3-031-17531-2_10

have also been used to find identical-by-descent (IBD) [5, 7, 16] segments shared among individuals of a population. IBD segments are segments of chromosome shared among individuals such that they share a most recent common ancestor. Numerous features of IBD segments including their counts, length distribution, etc. have been studied as they reveal useful information of the population history, selection pressure, and the disease loci [14]. Many other variations of the PBWT algorithm have been developed that tackle variety of problems. gPBWT [11] provides a method to efficiently query graph-encoded haplotypes, d-pbwt [13] provides efficient algorithms for query haplotype insertion and deletion, mPBWT [9] provides algorithms to deal with multi-allelic panels, [15] allows wildcard characters in the PBWT panel to study relative fitness of genomic variants, and cPBWT [10] and related works [1, 2] extend pairwise segment matching to multi-way matching, i.e., clusters of haplotype matches.

It is of interest to study composite haplotype matching patterns. For example, two long segment matches of a haplotype that are adjacent to each other may indicate a recombination event or an error of the phasing method. Another example is a combination of three segment matches, two long ones between the same pair of haplotypes, surrounding a relatively short one in the middle, that is a hallmark of a gene conversion. However, most existing problem formulations of PBWT algorithms are to find single matching segments between pairs of haplotypes or a single matching block among a cluster of haplotypes.

Recently, bi-directional PBWT [8] (or bi-PBWT) was the first to study composite matching patterns of more than one matching blocks. bi-PBWT finds all matches between sufficiently long matching blocks at both sides of a site, with a small gap of tolerance. The bi-PBWT algorithm is a two-pass scanning algorithm, first scanning backwards, storing the reverse PBWT data structure, and then a second pass scanning forward and makes the block matching. We wish to generalize the two-pattern matching problem to more, and with a more general definition of the connections between individual matching patterns.

Here, we formulate the problem of composite haplotype matching. Conceptually, the goal of composite haplotype matching problem is to find a number of pairwise matching segments or matching clusters, each one is long enough, with small enough gaps/overlaps, and the haplotype IDs of these segments satisfy certain condition (e.g., having a single id shared with all segments, and the other IDs may or may not belong to the same individual). The phasing error pattern, recombination, and the gene conversion each can be seen as special cases of such composite haplotype matching patterns.

In this paper, we introduce a space-efficient algorithm, mcPBWT, that utilizes two or more synchronized scans over multiple columns of PBWT to compare and analyze multiple sets of matches. While a naive solution in the style of bi-PBWT that stores pre-computed PBWT is time-efficient, the space-efficiency of pre-computed panel can be inconveniently large for biobank-scale data. Our algorithm's multi-column idea allows various columns of PBWT to exchange information and integrate multiple single matches without a large memory-footprint or disk-usage. The algorithm also makes a single pass of a haplotype panel. This online nature and generalizability of the algorithm will provide an efficient way of studying complex set of matches.

2   Preliminaries

2.1   PBWT Overview

PBWT (positional Burrows-Wheeler Transform) is an efficient data structure that finds all matches of user-specified minimum length (L) in an efficient manner given a panel of haplotype sequences. In his paper, Durbin [4] defines a *haplotype panel* X as a set of M haplotype sequences $x_i \in X$, $i = 0, 1, 2, ..., M - 1$. Each sequence x_i has N SNP sites indexed by k, $k \in \{0, 1, ...N - 1\}$. All the sites are assumed to be bi-allelic, namely $x_i[k] \in \{0, 1\}$. *Locally maximal match* is defined as a match between two haplotype sequences s and t from k_1 to k_2 such that, $s[k_1 - 1] \neq t[k_1 - 1]$, or $k_1 = 0$ and $s[k_2] \neq t[k_2]$, or $k_2 = N$. A match is a *long match* if it is locally maximal and its length satisfies a user-specified length threshold. *Prefix array* a contains $N + 1$ reverse prefix sorted orderings of the sequences, one for each $k \in 0...N$. It can also be thought of as a permutation of indices of the haplotype sequences that range from 0 to $M - 1$ for every $k \in \{0, 1, 2...N\}$. a_k is the k-th reverse sorted ordering of the haplotype sequences up to the site $k - 1$. In any a_k, adjacent sequences are maximally matching until k. y_i^k is the i-th sequence in a_k, $y_i^k = x_{a_k}[i]$. The divergence array d_k stores the starting position of locally maximal matches ending at k between a sequence and the preceding sequence in a_k.

2.2   Composite Haplotype Matching

Here, we generalize the notion of haplotype matching in a panel. A *single haplotype match pattern* (or single match) in a haplotype panel is defined as $p = (c, k_1, k_2)$, where c is a subset of the total set of haplotype indexes $C = \{0, ..., M - 1\}$, and the haplotype sequences match between sites k_1 and k_2: $x_i[k_1, k_2) = x_j[k_1, k_2)$, for any $i, j \in c$. Here, the *length* of the pattern is $L(p) = |k_2 - k_1|$, and the *width* of the pattern is $W(p) = |c|$. We can also denote the *sequence id set* of p as $c(p) = c$, the *left boundary* of p as $l(p) = k_1$, and the *right boundary* as $r(p) = k_2$. In general, $|c| \geq 2$ indicates a cluster of haplotypes matching. For pairwise matching, $|c| = 2$. The problem of single pattern haplotype matching is, given a predefined length cutoff L and width cutoff W, find all patterns p, such that $L(p) \geq L$ and $W(p) \geq W$ in a haplotype panel.

Further, for two single haplotype match patterns p and q, we define q *g-follows* p if they are adjacent, i.e., the *gap* (or *overlap*) between them, $g(p, q) = l(q) - r(p)$, is small: $|g(p, q)| \leq g$ and some haplotypes are shared among their sequence id sets $c(p) \cap c(q) \neq \varnothing$.

With that, we define a *composite haplotype match pattern* in a haplotype panel as a series of B single matches, $\mathcal{P} = \{p_b\}, b = 0...B - 1$, that g-follow each other, i.e., p_i g-follows p_{i-1}, for $i = 1..B - 1$, and they share some common haplotypes $c(\mathcal{P}) = \bigcap_{b=0}^{B-1} c(p_b) \neq \varnothing$. We call B the *span* of $\mathcal{P}$, and $c(\mathcal{P})$ the *thread* of $\mathcal{P}$.

The problem of composite haplotype matching pattern is, given a predefined set of length cutoffs $\{L_b\}$, $b = 0...B - 1$, width cutoff W, gap tolerance g, the span B, the thread width w, find all composite patterns $\mathcal{P} = \{p_b\}$, $b = 0..B - 1$ such that $L(p_b) \geq L_b$, $W(p_b) \geq W$, p_i g-follows p_{i-1}, for $i = 1..B - 1$, and $|c(\mathcal{P})| = w$. Of course, it is possible to specify different width cutoffs and gap tolerances for individual single match. We omit that for simplicity of presentation.

In this work, we mainly focus on *double haplotype match patterns* ($B = 2$) or *triple haplotype match patterns* ($B = 3$). We will also mainly focus on pair segment matching ($W = 2$), and single thread composite patterns ($w = 1$). We will present memory-efficient multi-column scanning PBWT-based algorithms: *double-PBWT* for identifying double haplotype match patterns, and *triple-PBWT* for triple haplotype match patterns.

3   Multi-column PBWT

In PBWT, a single column scans the haplotype panel from left to right and updates divergence values and prefix arrays to output long matches. Let $P_{k,L}$ signify the active single column of PBWT operating at site k finding matches of length at least L **ending** at site k. If the panel were to be scanned in a reverse fashion from right to left then, $R_{N-k,L}$ would signify the active single column of PBWT operating at site $N - k$ finding matches of length at least L ending at site $N - k$. This would allow the user to compare two sets of matches at site k (or $N - k$ if looking from right to left) i.e. matches from $P_{k,L}$ and $R_{N-k,L}$. This method would require the panel to be scanned twice if we wanted to compare such matches along different SNP sites of the panel. In fact, one approach would be to scan the panel in reverse and save all the precomputed divergence values and prefix arrays before hand as in [8]. While this method works well, it occupies disk-space and needs to be loaded into memory which might not be efficient for larger panels. In contrast, mcPBWT is capable of providing the same information on a single pass of the haplotype panel from left to right using multiple columns of PBWT without having to pre-scan the panel and save the values on disk.

The general idea of mcPBWT is to utilize the information obtained from "look-ahead" PBWT columns. Here, a new PBWT column operating at site $k + L$ that finds all the matches **starting** at site k is denoted by $P'_{k+L,L}$ such that matches found by columns P' and R are the same i.e. $P'_{k+L,L} = R_{N-k,L}$. So mcPBWT would consist of a set of PBWT columns where the columns are finding matches ending at a site or starting few sites before. Even though, we only talk about two columns of PBWT here, similar approach can be also used for the case of three columns of PBWT. In fact, we can spawn multiple such columns of PBWT and at each column, the user has the flexibility to find either matches ending at that site as in PBWT or have the flexibility to find matches starting certain number of sites before, depending on the use case. First, we discuss the divergence value properties that enable us to find the matches **starting** at a certain site and then the two specific cases of mcPBWT in the following sections.

3.1   Divergence Value Properties

There are two major properties of the divergence values that assist in finding the matches starting at a certain site efficiently. The first property asserts that in a divergence array, adjacent divergence values cannot be equal unless it is zero (or when a match does not exist) [4], while the second property that extends the first property asserts that between any two consecutive equal divergence values there must be a divergence value that is greater than those equal values. These properties are presented formally as two lemmas below.

Lemma 1. *Two adjacent divergence values aren't equal unless it is zero (or when the divergence value greater than current k, i.e. when there is no match).*

Proof. This property mentioned by Durbin, asserts that $d_k[i-1] \neq d_k[i]$, $0 < i < M$ except when $d_k[i] = 0$ or $d_k[i] = k$. For any index i, the divergence value at some site k, $d_k[i]$ gives the starting position of a match between haplotypes at index i and $i-1$. Since the panel is bi-allelic and reverse prefix sorted, this means that $y_{i-1}[d_k[i]-1]$ must be 0 and $y_i[d_k[i]-1]$ must be 1. The same condition holds for $d_k[i-1]$ in that $y_{i-2}[d_k[i-1]-1]$ must be 0 and $y_{i-1}[d_k[i-1]-1]$ must be 1. But if we assume that $d_k[i] = d_k[i-1]$, there is a contradiction on the value for $y_{i-1}[d_k[i]-1]$. This proves that the adjacent divergence values can't being equal unless the divergence value is equal to 0 or k (match does not exist).

Lemma 2. *In a divergence array d_k, assume $d_k[i] = x$, and there exists another index $g, g < i$ and $g \neq i-1$ where g is the first index preceding i to have divergence value equal to x i.e. $d_k[g] = x$, then there must be a divergence value greater than x at index h, where $g < h < i$ and $d_k[h] > x$ (except when $x = 0$) for $0 < g, h, i < M$.*

Proof. This property is essentially an extension of Lemma 1. In a bi-allelic panel, $d_k[i] = x$ asserts that $y_{i-1}[x-1] = 0$ and $y_i[x-1] = 1$. For index h in the range (g, i), the divergence value can either be $d_k[h] < x$ or $d_k[h] > x$ since g is the first index preceding i where $d_k[g] = x$. If we assume that all the divergence values in the range (g, i) are less than x, we can conclude that $y_h[x-1] = 0$, $\forall\ g < h < i$. This includes $d_k[g+1] < x$, which implies that $y_{g+1}[x-1] = y_g[x-1] = 0$. However, we already have the case that $d_k[g] = x$, which means that $y_g[x-1] = 1$ and $y_{g-1}[x-1] = 0$. That is a contradiction for the value of $y_g[x-1]$ which proves that it cannot be the case that all the divergence values are less than x. Hence, there must be a divergence value greater than x in the range (g, i).

3.2   Finding Blocks of Starting Matches

In a PBWT panel, neighboring haplotypes sharing matching segments cluster together. Such a collection of neighboring matches is called a *block*. Such blocks of

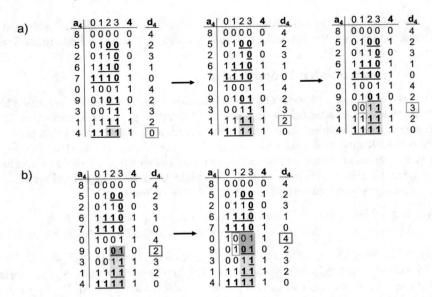

Fig. 1. Two blocks of matches of minimum length 2 starting at site $k = 2$ in a panel of 10 haplotypes reverse prefix sorted at $k = 4$. The process of block detection while scanning the divergence array is shown. The rightmost colored rectangle (under d_4) shows the divergence value being scanned while the solid colored rectangles, blue in (a) and red in (b), show the actual matching blocks. The grey boxes at position $k = 1$ show the two groups within each block.

matches are separated by a haplotype whose divergence value is greater than the difference of the site being observed and the length threshold specified. When iterating over the divergence array at a certain site, d_k, the divergence value properties discussed above restrict the combinations for the ordering of divergence values in the array. This in-turn assists in finding such blocks of matches where the matches share the same starting position.

Figure 1 shows an example of using the divergence value properties to find blocks of starting matches of length, $L \geq 2$. It shows a reverse prefix sorted panel of 10 haplotypes at site $k = 4$, where d_4 shows the values of the divergence array and a_4 shows the prefix array. Figure 1(a) shows the first block of starting matches found and Fig. 1(b) shows the second block found. These two blocks are separated by the sequence indexed 3 in a_4 with divergence value 3 which is greater than $k - L = 2$. The grey box at site $k = 1$ in the figures show the two clusters being formed within a block, called *groups*. Within each block, the top group consists of zeros at site $k - L - 1$ and the bottom group consists of ones at that site. It's also to be noted that a block of starting matches must contain only one sequence with divergence value equal to $k - L$ value. In the blue block, that sequence is 1, and in the red block, that sequence is 9. The haplotypes belonging to the same block but different groups constitute the actual matches that are of

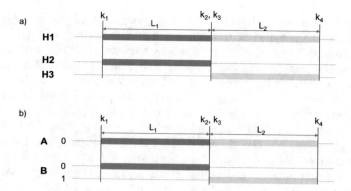

Fig. 2. Double haplotype match patterns (a) Double haplotype match pattern where $c(p_0) = \{H1, H2\}$, $c(p_1) = \{H1, H3\}$ and $L(p_0) \geq L_1$, $L(p_1) \geq L_2$. Such formation represents crossover breakpoints. (b) Alternating match where A and B are two individuals and individual B has an alternating match with respect to individual A.

minimum length 2 and start at site $k = 2$. These starting matches are given by the tuples $(3, 1), (3, 4)$ for the blue block and $(0, 9)$ for the red block.

3.3 double-PBWT

double-PBWT, as the name suggests, uses two columns of PBWT simultaneously scanning the panel from left to right. Let D_{k,L_1} and D'_{k',L_2} be the two columns where column D_{k,L_1} is operating at site k finding matches of length at least L_1 ending at site k while column D'_{k',L_2} is operating at site k', L_2 sites ahead of k such that $k' = k + L_2$. This leading column finds matches of length at least L_2 starting from site k. Such a formulation of mcPBWT enables us to evaluate matches flanking on either side of a site. Such flanking matches can be defined by a composite haploytpe match pattern with $B = 2$, namely, double haplotype match pattern where $\mathcal{P} = \{p_0, p_1\}$, $|c(\mathcal{P})| = 1$, $W(p_0) = W(p_1) = 2$ and $g = 0$(for simplicity). This form of composite pattern is structured to signify patterns of recombination. Such combination of matching pair segments were also shown to be potential phasing errors [5]. Here, we introduce and focus on one such variation of this composite pattern termed alternating match but the algorithm can find all the other possible variations as well. We define alternating match and the problem statement formally in the following paragraphs.

Alternating Match Definition. An alternating match is a strict case of a double haplotype match pattern, $\mathcal{P} = \{p_0, p_1\}$, $|c(\mathcal{P})| = 1$, $W(p_0) = W(p_1) = 2$ where individual information is also encoded. If we assume the single thread haplotype belongs to an individual say, A, the non-thread haplotypes in p_0 and p_1 must be the complementary haplotypes of the same individual, say, B. Such a case of double haplotype match pattern is termed as an *alternating match*. For instance, if p_0 is a match between $A_i[k_1, k_2)$ and $B_j[k_1, k_2)$where, $i, j \in \{0, 1\}$ and A_i, B_j are haplotypes of individuals A and B respectively and A_i is the thread haplotype. Then p_1 must be a match between $A_i[k_3, k_4)$ and $B_{j'}[k_3, k_4)$, where $B_{j'}$ is the complement haplotype of individual B. Here, B is said to have an alternating match with respect to A. Figure 2 shows the double haplotype match pattern and an alternating match. Figure 2(a) shows the double haplotype match pattern where $H1, H2$ and $H3$ are three haplotypes. Here, sequence id set of p_0, i.e., $c(p_0) = \{H1, H2\}$ and sequence id set of p_1, $c(p_1) = \{H1, H3\}$ where $c(p_0) \cap c(p_1) = \{H1\}$ is the thread haplotype. This form of double haplotype match represents crossover breakpoints. Similarly, Fig. 2(b) shows an example of an alternating match where individual B has an alternating match with respect to individual A. Here, sequence id set of p_0, i.e., $c(p_0) = \{A_0, B_0\}$ and sequence id set of p_1, $c(p_1) = \{A_0, B_1\}$ such that A_0 is the thread haplotype and B_0 and B_1 are complementary haplotypes of individual B. The first segment of the alternating match (the blue segment) is of length L_1 and the second yellow segment is of length L_2. Figure 2(b) shows the case where the boundaries of the two pairs are juxtaposed such that $k_2 = k_3$, i.e. $g = 0$.

Problem Statement. Given a phased haplotype panel and user specified length parameters L_1, L_2 and g (here assumed 0 for simplicity), find all the alternating matches.

We discuss the simple case of the alternating matches with strict boundaries as shown in Fig. 2(b) such that $g = 0$. The algorithm proceeds where the leading PBWT column D' first stores the information on starting matches using block and group properties, passes this information to the lagging PBWT column D which then checks if an alternating match exists for each ending match pair found of minimum length L_1. Since the two PBWT columns are L_2 ($k' - k = L_2$) distance apart the leading PBWT column finds matches at least L_2 long starting at k.

Algorithm 1 shows the working mechanism for PBWT column D'. It updates the block and group information for all the starting matches found at a given site. The *block* array of size M keeps track of the block-membership of all the haplotypes, where the index of the array represents the haplotypes' indexes. An integer value (id) is assigned to haplotypes belonging to the same block. *group*, an array of size M distinguishes between the haplotypes of the block with 0 or 1 at $k' - L_2 - 1$ position. The index of this array also represents the haplotypes. *rblock* stores the same block and group information in a dictionary format where it is indexed by the block id to find the double haplotype match pattern show in Fig. 2(a). The divergence values and prefix arrays are calculated using Durbin's algorithm 1 and 2 [4]. The *block* and *group* arrays along with *rblock* are updated simultaneously as the divergence and prefix arrays are updated so the time complexity to find the blocks of starting matches is $O(M)$ at each site and $O(MN)$ for the all sites. The block and group arrays are passed on to the PBWT column D where it decides if an alternating match is found. Here, the divergence array is scanned from $M - 1$ to 0 as it's more intuitive to understand

Algorithm 1. Leading PBWT at column D'_{k',L_2}: Find blocks and groups of matches starting at position k

```
1   create arrays block[], group[], s[]
2   create rblock{}                          ▷ Store haplotypes belonging to blocks
3   id ← 1
4   f ← true
5   for i = M − 1 to 0 do
6     if f then                              ▷ Start looking for a block
7       if d_{k'}[i] < k' − L_2 then
8         s.add(a_{k'}[i])
9       else if d_{k'}[i] == k' − L_2 then    ▷ Block exists
10        s.add(a_{k'}[i])
11        create rblock{id} ← {}
12        for all j ∈ s do
13          block[j] ← id
14          group[j] ← 1
15          rblock{id}.add(j)
16        f ← false, clear s
17      else
18        clear s
19    else
20      if d_{k'}[i] < k' − L_2 then
21        s.add(a_{k'}[i])
22      else                                  ▷ Block ends
23        s.add(a_{k'}[i])
24        for all j ∈ s do
25          block[j] ← id
26          group[j] ← 0
27          rblock{id}.add(j)
28        id ← id + 1, f ← true
```

the formation of block and groups but it can be scanned in the other direction without affecting the algorithm as shown later in Algorithm 4

Algorithm 2 is responsible for finding the other set of matches ending at site k and deciding if an alternating match exists. This algorithm scans the panel from site 0 to N simultaneously as D' scans the panel ahead of it. It finds all matches ending at site k using Durbin's algorithm 3 [4]. For every such ending match, it checks to see if there's a starting match that satisfies as an alternating match. This is done using the block and group arrays passed from D'. When the condition is met, the alternating match is reported. This checking is done in constant time since array access is constant time. Because of this constant time lookup for every pair of ending match found, the time complexity depends on the number of such match pairs found. We define C as the total number of match pairs found across all the sites. Therefore, the time complexity for this column is $O(C)$.

Algorithm 2. Lagging PBWT at column D_{k,L_1}: Report matches of length at least L_1 forming alternating matches

1 block [], group [], rblock{}	▷ Obtained from Algorithm 1
2 $u \leftarrow 0$, $v \leftarrow 0$, create empty arrays a[],b[]	▷ Modified Durbin's algorithm 3

```
 3  for i = 0 to M − 1 do
 4      if d_k[i] > k − L_1 then
 5          if u > 0 and v > 0 then
 6              for all 0 ≤ i_u < u and 0 ≤ i_v < v  do
 7                  for match from a[i_u] to b[i_v] ending at k_b do
 8                      if block[a[i_u]] = block[b[i_v]^c] then    ▷ b[i_v]^c is the complement of
```
$b[i_v]$
```
 9                          if group[a[i_u]] ≠ group[b[i_v]^c] then
10                              report alternating match
11                      if block[b[i_v]] = block[a[i_u]^c] then
12                          if group[b[i_v]] ≠ group[a[i_u]^c] then
13                              report alternating match
14          u ← 0, v ← 0
15      if y_i[k] = 0 then
16          a[u] ← a_k[i], u ← u + 1
17      else
18          b[v] ← a_k[i], v ← v + 1
```

Algorithm 3. *double-PBWT*: Simultaneous run of two PBWT columns

```
 1  k ← 0, k' ← 0
 2  while k' < N do
 3      run Algorithm 1                         ▷ compute block[] and group[] for k'
 4      if k' − k ≥ L_2 then                    ▷ feed block[] and group[] to algorithm 2
 5          run Algorithm 2                     ▷ report alternating matches at k
 6          k ← k + 1
 7      k' ← k' + 1
```

It is to be noted that this algorithm does not handle for the double haplotype match pattern in Fig. 2(b) but *rblock* can be used to query such patterns easily. For every ending match pair, like $H1$ and $H2$ detected by lagging PBWT at column D, the haplotypes belonging in the blocks of $H1$ and $H2$ are scanned using *block* and *rblock*, to find the second matching segment $H1$ and $H3$ ($H1$ being the thread haplotype) or $H2$ and $H3$ ($H2$ being the thread haplotype). This scanning process is done in $O(b)$ time for every ending match pair where b is the average number of haplotypes in a block. Hence, the overall time complexity for such an algorithm would be $O(C * b)$ across all the sites.

Algorithm 3 shows the synchronous execution of both PBWT columns as it does a one pass scan on the panel. Since both columns move simultaneously across all the sites, the overall complexity of the algorithm is $O(MN+C)$. While we only show the case of alternating match when $g = 0$, these algorithms can be extended to handle for overlaps or gaps ($g \geq 1$) by altering the distance between the two PBWT columns.

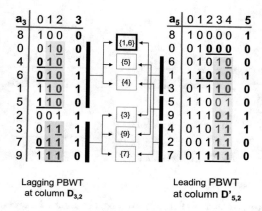

Lagging PBWT
at column $D_{3,2}$

Leading PBWT
at column $D'_{5,2}$

Fig. 3. Double PBWT finding blocks of matches of length $L_1, L_2 \geq 2$ and $W' \geq 2$. (Left) Haplotype panel with 10 haplotypes reverse prefix sorted at $k = 3$. The two colored boxes represent the block of matches ending at $k = 3$. (Right) Haplotype panel reverse prefix sorted at $k' = 5$ where the colored boxes represent blocks of matches starting at $k = 3$. (Middle) The boxes in the middle show haplotypes common between the lagging and leading blocks. The grey block from lagging PBWT column $D_{3,2}$ and green block from leading PBWT column $D'_{5,2}$ share two haplotypes $\{1, 6\}$ to satisfy $W' \geq 2$ showing that the two haplotypes have an extending match.

3.4   double-PBWT: Comparing Block of Matches

So far we've used *double-PBWT* to find double haplotype composite matching patterns where $W(p_b) = 2$, $b = 0, 1$ but here we take advantage of its versatility to make comparisons between blocks of matches, i.e. $W(p_b) \geq 2$, $b = 0, 1$. The main idea here is to only evaluate matching blocks found by the two PBWT columns when they satisfy user-specified constraints of a valid block structure. A block structure is defined as a block consisting of at least W' haplotypes in common and sharing at least L long segments. This definition is adapted from cPBWT [10] and allows us to process composite match patterns in blocks. Such block-based comparison can be useful in studying recombination patterns too [8]. Here, both columns of PBWT store haplotypes that belong to different blocks and the blocks found by the two columns are compared to see if they share at least W' haplotypes. When this requirement is met, the blocks are output. While Fig. 3 shows the comparison of blocks of haplotypes, this can also be extended to find alternating matches. Since alternating matches have more structure in terms of individuals that the haplotypes belong to, the algorithm needs to be modified to account for this constraint. For alternating matches in a block structure, additional constraints can be specified for the minimum number of thread haplotypes i.e. $|c(\mathcal{P})| \geq w_{min}$ and the minimum number of alternating individuals that should be present in a block-structure.

Figure 3 shows an example of analyzing blocks of matches of length $L_1, L_2 \geq 2$ on either side of site 3. Here, a valid block structure should have at least

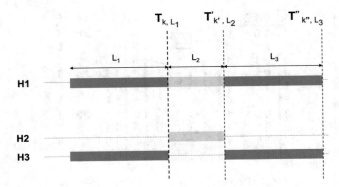

Fig. 4. A composite pattern of three matching segments between three haplotypes $H1, H2$ and $H3$. This pattern is representative of gene conversion, where the yellow segment is a possible gene conversion tract such that $L_2 << L_1, L_3$. The vertical dashed lines show the three simultaneous PBWT runs each operating at sites k, k' and k'', where $k' = k + L_2$ and $k'' = k + L_2 + L_3$.

2 haplotypes ($W' \geq 2$). The only valid block structure found is shown in the middle with two haplotypes $(6, 1)$ common to the two top blocks indicating that they share an extending match. It can be seen that this can be generalized to handle mismatches to study recombination patterns by adjusting the distance between the two PBWT columns.

3.5   triple-PBWT

triple-PBWT is the case of mcPBWT where three columns of PBWT are utilized. Each column has the freedom to find matches ending at those sites or starting few sites before. Here, we define *triple-PBWT* with three columns T_{k,L_1}, T'_{k',L_2} and T''_{k'',L_3} where, $k' = k + L_2$ and $k'' = k + L_2 + L_3$.

triple-PBWT can be useful in finding a triple haplotype composite match pattern $(B = 3)$, $\mathcal{P} = \{p_b\}$, $|c(\mathcal{P})| = 1$ and $W(p_b) = 2, b = 0...2$. The set of length constraints for the three matching pairs are $\{L_1, L_2, L_3\}$ such that $L(p_0) \geq L_1, L(p_1) = L_2$ and $L(p_2) \geq L_3$ and $L_2 << L_1, L_3$. Here, L_2 is restricted to be a short match in comparison to L_1 and L_3 to emulate a gene-conversion tract. Figure 4 shows an example of triple haplotype match pattern. Here, $H1, H2$ and $H3$ are three haplotypes where, $H1$ is the thread haplotype. Similarly, the sequence id sets are $c(p_0) = \{H1, H3\}$, $c(p_1) = \{H1, H2\}$ and $c(p_2) = \{H1, H3\}$. The three columns of *triple*-PBWT are represented by the vertical dashed lines. Here, column T runs a standard PBWT finding matches of length at least L_1 ending at site k. The T' column finds matches of exact length L_2 starting at site k and ending at site $k + L_2$ and column T'' finds matches of length at least L_3 starting from site $k + L_2$. While a simple composite pattern like the one shown can be representative of gene-conversion tract, it's not sufficient condition and care has to be taken since the smaller match pair p_1 could end up providing

Algorithm 4. *triple-PBWT* at column T': Find matches that are of exact length L_2 at site k'

```
1  create block[], group[], s[]                    ▷ For queries of exact match pairs
2  create rblock{}                                 ▷ Store haplotypes belonging to blocks
3  Create end[]                        ▷ Tracks haplotypes that have 0 or 1 in the next site
4  id ← 1, f ← False
5  for i = 0 to M − 1 do
6      if d_{k'}[i] > k' − L_2 then
7          if not s.empty() and f then
8              for e ∈ s do
9                  block[e] ← id, group[e] ← 1
10                 rblock{id}.add(e)
11             id ← id + 1
12             s.clear(), s.add(a_{k'}[i])
13             f ← False
14         else
15             s.clear(), s.add(a_{k'}[i])
16     else if d_{k'}[i] < k' − L_2 then
17         s.add(a_{k'}[i])
18     else if d_{k'}[i] == k' − L_2 then
19         f ← True
20         if not s.empty() then
21             Create rblock{id} ← {}
22             for e ∈ s do
23                 block[e] ← id, group[e] ← 0
24                 rblock{id}.add(e)
25             s.clear(), s.add(a_{k'}[i])
26     if y_i[k'] = 0 then
27         end[i] ← 1                                      ▷ haplotype ends in 0
28     else
29         end[i] ← −1                                     ▷ haplotype ends in 1
30 if not s.empty() and f then                            ▷ Boundary case
31     for e ∈ s do
32         block[e] ← id, group[e] ← 1
33         rblock{id}.add(e)
```

lots of false positives. Additional information has to be incorporated to this formulation to distinguish false positives from true gene-conversion tracts but this shows one potential use case for the algorithm. The first column behaves similar to Algorithm 1 and can be easily extended from there. The last column runs Algorithm 2 as in *double-PBWT*. The only change is for PBWT column T'. Algorithm 4 shows how the exact matches can be catalogued for column T'. Exact matches satisfy both restrictions of starting and ending matches and hence the algorithm uses ideas of both starting and ending matches to find them. Here, *block* and *group* arrays along with *rblock* serve the same function as in *double-*

PBWT. rblock is a dictionary indexed by the block ids that store haplotype indices belonging to such blocks. This dictionary is used to find exact match pairs like p_1. A new array *end* is introduced which keeps track of whether the haplotypes have 0 or 1 in the next variant site. This *end* array is utilized to filter the exact matches from the starting matches. Since, the *end* array is updated along with divergence and prefix arrays, the time complexity for this algorithm at a given site is $O(M)$. The overall synchronization of the three columns is similar to *double-PBWT* in that columns T' and T'' catalogue the exact matches and starting matches respectively and pass this information in the form of *block*, *group*, *rblock* and *end* (for column T') to column T. Then, for every ending match pair detected by column T, it takes constant time to look up match pairs p_2 but the haplotypes that belong to $H1$'s block have to be scanned for match pair p_1. When those matching pair segments exist, the triple haplotype match is reported.

Of the three PBWT columns of *triple-PBWT*, column T'' is the same as the leading column of *double-PBWT* and hence has a time complexity of $O(M)$ at each site and $O(MN)$ across all variant sites. For the middle column T', the time complexity to catalogue the exact matches is $O(M)$ at each site and $O(MN)$ across all the sites. It's important to note that querying of exact matches of length L_2 can be done by accessing *block*, *group*, *end* arrays in constant time. Lastly, column T makes constant time query for every ending match found to see if the last match pair exists but has a time complexity of $O(b)$ to find the middle match pair, where b is average number of haplotypes in a block. Since, T is our trigger column, the time complexity depends on the number of match pairs found by PBWT column T. Hence, we define the complexity of this column across all sites similar to *double-PBWT* as $O(C*b)$, where C is the total number of match pairs found across all the variant sites. The overall time complexity of Triple PBWT is then $O(C*b + MN + MN)$, i.e. $O(MN + C*b)$.

4   Discussion

In this work, we present a more flexible and powerful variation of PBWT for detecting composite haplotype matches. The original formulation in PBWT for the haplotype matching problem only captures the matching pattern at every single column separately. Our algorithm, however, simultaneously captures the patterns across multiple columns of PBWT. In the single column matching formulation, at the active column of the PBWT, one only has access to the information in the past but is uninformed about future columns. In our formulation, the columns at the forefront can provide "look ahead" information allowing the algorithm to make complex decisions. Our flexible algorithm's capability to analyze composite matching patterns opens new potentials of the PBWT data structure.

The proposed method does not require to output or book keep the matches which will be very useful in analyzing large haplotype panels with millions of individuals. While the PBWT algorithm is able to find all matches efficiently, the number of matches in large cohorts may be enormous. As a result, analyzing composite patterns like alternating matches after the PBWT run may not be very efficient. Hence, a flexible algorithm like *double-PBWT* would be useful in such cases. Additionally, we also showed that the *triple-PBWT* could find composite haplotype match representative of gene-conversion tracts. This shows that mcPBWT has potential to allow and adjust for flexible matching criteria and is suitable for more general-purpose settings. The double haplotype match pattern discussed where $H1$ is the thread haplotype and $H2$ and $H3$ have matches with it not only identifies more recombination events but also provides plausible evidence that $H2$ and $H3$ coalesce more recently. This could help to determine the time of the recombination events, and also help "triangulating" the genealogical relationship among individuals carrying these matching segments. Such analyses can also be conducted using blocks to enable stronger signal using mcPBWT.

Acknowledgments. PS, AN, DZ and SZ were supported by the National Institutes of Health grant R01 HG010086. AN, DZ and SZ were also supported by the National Institutes of Health grants R56 HG011509. AN and DZ were also supported by the National Institutes of Health grant OT2-OD002751.

References

1. Alanko, J., Bannai, H., Cazaux, B., Peterlongo, P., Stoye, J.: Finding all maximal perfect haplotype blocks in linear time. Algorithms Mol. Biol. **15**(1), 1–7 (2020)
2. Cunha, L., Diekmann, Y., Kowada, L., Stoye, J.: Identifying maximal perfect haplotype blocks. In: Alves, R. (ed.) BSB 2018. LNCS, vol. 11228, pp. 26–37. Springer, Cham (2018). https://doi.org/10.1007/978-3-030-01722-4_3
3. Delaneau, O., Zagury, J.-F., Robinson, M., Marchini, J., Dermitzakis, E.: Accurate, scalable and integrative haplotype estimation. Nat. Commun. **10**(1), 1–10 (2019)
4. Durbin, R.: Efficient haplotype matching and storage using the positional Burrows-Wheeler transform (PBWT). Bioinformatics **30**(9), 1266–1272 (2014)
5. Freyman, W., et al.: Fast and robust identity-by-descent inference with the templated positional Burrows-Wheeler transform. Mol. Biol. Evol. **38**(5), 2131–2151 (2021)
6. Loh, P.-R., et al.: Reference-based phasing using the haplotype reference consortium panel. Nat. Genet. **48**(11), 1443 (2016)
7. Naseri, A., Liu, X., Tang, K., Zhang, S., Zhi, D.: RaPID: ultra-fast, powerful, and accurate detection of segments identical by descent (IBD) in biobank-scale cohorts. Genome Biol. **20**(1), 143 (2019)
8. Naseri, A., Yue, W., Zhang, S., Zhi, D.: Efficient haplotype block matching in bidirectional PBWT. In: Carbone, A., El-Kebir, M. (eds.) 21st International Workshop on Algorithms in Bioinformatics (WABI 2021). Leibniz International Proceedings in Informatics (LIPIcs), Dagstuhl, Germany, vol. 201, pp. 19:1–19:13. Schloss Dagstuhl - Leibniz-Zentrum für Informatik (2021)
9. Naseri, A., Zhi, D., Zhang, S.: Multi-allelic positional Burrows-Wheeler transform. BMC Bioinform. **20**(11), 1–8 (2019)
10. Naseri, A., Zhi, D., Zhang, S.: Discovery of runs-of-homozygosity diplotype clusters and their associations with diseases in UK biobank. medRxiv (2020). https://doi.org/10.1101/2020.10.26.20220004

11. Novak, A., Garrison, E., Paten, B.: A graph extension of the positional Burrows-Wheeler transform and its applications. Algorithms Mol. Biol. **12**(1), 1–12 (2017)

12. Rubinacci, S., Delaneau, O., Marchini, J.: Genotype imputation using the positional Burrows Wheeler transform. PLoS Genet. **16**(11), e1009049 (2020)

13. Sanaullah, A., Zhi, D., Zhang, S.: d-PBWT: dynamic positional Burrows-Wheeler transform. Bioinformatics **37**(16), 2390–2397 (2021)

14. Thompson, E.: Identity by descent: variation in meiosis, across genomes, and in populations. Genetics **194**(2), 301–326 (2013)

15. Williams, L., Mumey, B.: Maximal perfect haplotype blocks with wildcards. iScience **23**(6), 101149 (2020)

16. Zhou, Y., Browning, S.R., Browning, B.L.: A fast and simple method for detecting identity-by-descent segments in large-scale data. Am. J. Hum. Genet. **106**(4), 426–437 (2020)

Computational Advances in Molecular Epidemiology

Clustering SARS-CoV-2 Variants from Raw High-Throughput Sequencing Reads Data

Prakash Chourasia[1] , Sarwan Ali[1] , Simone Ciccolella[2] ,
Gianluca Della Vedova[2] , and Murray Patterson[1(✉)]

[1] Georgia State University, Atlanta, GA 30303, USA
{pchourasia1,sali85}@student.gsu.edu, mpatterson30@gsu.edu
[2] University of Milano-Bicocca, Milan, Italy
{simone.ciccolella,gianluca.dellavedova}@unimib.it

Abstract. The massive amount of genomic data appearing over the past two years for SARS-CoV-2 has challenged traditional methods for studying the dynamics of the COVID-19 pandemic. As a result, new methods, such as the Pangolin tool, have appeared which can scale to the millions of samples of SARS-CoV-2 currently available. Such a tool is tailored to take assembled, aligned and curated full-length sequences, such as those provided by GISAID, as input. As high-throughput sequencing technologies continue to advance, such assembly, alignment and curation may become a bottleneck, creating a need for methods which can process raw sequencing reads directly.

In this paper, we propose several alignment-free embedding approaches, which can generate a fixed-length feature vector representation directly from the raw sequencing reads, without the need for assembly. Moreover, because such an embedding is a numerical representation, it can be passed to already highly optimized clustering methods such as k-means. We show that the clusterings we obtain with the proposed embeddings are more suited to this setting than the Pangolin tool, based on several internal clustering evaluation metrics. Moreover, we show that a disproportionate number of positions in the spike region of the SARS-CoV-2 genome are informing such clusterings (in terms of information gain), which is consistent with current biological knowledge of SARS-CoV-2.

Keywords: Clustering · SARS-CoV-2 · High-throughput sequencing · Assembly · Alignment-free

1 Introduction

COVID-19 is the first major global pandemic that has occurred during an era of the widespread availability of high-throughput sequencing technologies [48]. As a result, the amount of sequencing data for SARS-CoV-2 is orders of magnitude greater than for any virus in history. Worldwide efforts in the collection,

M. S. Bansal et al. (Eds.): ICCABS 2021, LNBI 13254, pp. 133–148, 2022.
https://doi.org/10.1007/978-3-031-17531-2_11

assembly, alignment and curation of high-throughput sequencing data such as GISAID [20] have greatly facilitated research efforts in producing publicly available high quality full-length nucleotide and amino acid (*e.g.*, spike) sequences in easy-to-use formats (*e.g.*, FASTA). The number of sequences on GISAID is more than 7 million to date. Because publishing a sequence on GISAID takes time, resources, and quality control, there is likely an even larger volume of raw high-throughput sequencing reads samples which have yet to be assembled or aligned. This amount will only increase as countries across the world begin to invest heavily in sequencing infrastructure for monitoring COVID-19 as well as future pandemics [34,39]. The need for agile data analysis procedures which can quickly and automatically fathom volumes of such data, possibly in real-time, will hence be crucial in this monitoring effort. Such procedures may not have time to perform genome (or proteome) assembly or alignment, which could become the bottleneck since they tend to be costly operations from a computational standpoint [45]. Moreover, the choice and parameterization of assembler usually requires some expert knowledge, and nonetheless results in an inherent bias that we would like to remove [21].

By the end of 2020, the number of curated full-length sequences available on databases such as GISAID neared one million [20], placing the analysis of such sequences—to retrieve any information about the evolution, diversity and dynamics of this virus—into big data territory. This quickly rendered obsolete many of the methods traditionally used for studying viruses, which are based on constructing a phylogenetic tree, such as Nextstrain [22], which can scale to thousands of sequences at most. Even state-of-the-art phylogenetic tree construction methods such as IQ-TREE 2 [31] can process tens of thousands of sequences at most, by making use of parallelization. The current methods which can scale to millions of sequences employ some form of clustering or classification, either as an alternative to or as a way to boost phylogenetic reconstruction. These methods tend to cluster or classify certain properties of (*e.g.*, SARS-CoV-2) sequences, one of the main properties being lineage label (*e.g.*, B.1.1.7 or the Alpha variant). From some of the foremost phylogenetics studies on the lineage dynamics of SARS-CoV-2, *e.g.*, [15], was a machine learning (ML) framework built on top to result in the state-of-the-art Pangolin tool classifier [37]. The Pangolin tool is currently what is used to build metadata, such as lineage labels for GISAID sequences. Other methods, still based on clustering and classification, but developed completely separately from the Pangolin tool and its body of literature, have also appeared [5,7,8,30]. Such methods have been shown to possess good predictive power on sets of GISAID sequences with "ground truth" lineage labels—assigned by the state-of-the-art Pangolin tool. Some of these methods have already been demonstrated to scale to a million [30] or several million [8] sequences.

These methods based on clustering and classification have addressed the scalability issue. However, they have only been tested and validated on these curated full-length nucleotides or amino acid sequences from, *e.g.*, GISAID. In light of the inundation of unassembled, unaligned raw high-throughput sequencing reads data that is to come in the near future, we explore in this paper how these existing tools can be adapted to this setting. The state-of-the-art Pangolin tool is designed to take full-length SARS-COV-2 nucleotide sequences as input, making a direct adaptation to the high-throughput reads setting difficult. Interestingly, however, the methods [5,8], based on k-mers, are *alignment-free* by nature, allowing their direct application to such raw high-throughput sequencing reads. In this paper, we apply these alignment-free approaches, including another alignment-free approach we design here based on the notion of *minimizer*, to a set of $\approx$7K high-throughput SARS-CoV-2 reads samples from PCR tests. We assess these approaches based on how they *cluster* this set, using several internal clustering quality metrics. We also compare approaches to each other using comparison metrics such as the adjusted Rand index (ARI). Because the Pangolin tool is considered to be the state-of-the-art, we also align each sample to the SARS-CoV-2 reference genome to extract a consensus sequence in order to allow to obtain a clustering with the Pangolin tool, for comparison purposes. This also allows to show that the genomic positions in these aligned sequences which contribute the most information (in terms of information gain) to the alignment-free clusterings are consistent with known properties of the SARS-CoV-2 genome. Finally, we perform various analyses to understand the statistical properties of our alignment-free embedding, such as compactness.

The rest of the paper is organized as follows. Section 2 contains related work. Section 3 describes the generation of a consensus sequence from a reads sample, the alignment-free approaches we propose, as well as the clustering methods we use, including the Pangolin tool as a baseline. Section 4 contains our experimental setup, including some dataset statistics. Section 5 details and discusses the results of these experiments. Finally, Sect. 6 concludes this paper, and highlights some potential avenues for future work.

2 Related Work

In recent years, some effort has been made to understand the behavior of SARS-CoV-2 using machine learning models by applying classification and clustering on the protein sequences [5,7,9,47,49]. Although these studies involve k-mers to generate fixed length feature embeddings that can be used as input to machine learning models, it is not clear if these methods would work as well on the raw sequencing reads samples directly, since they have only been proven on full-length spike or nucleotide sequences. Some effort has been made to classify metagenomic data [25,53]. However, it is unclear if those methods could be applied to SARS-CoV-2 reads samples. Therefore, there is a need to explore the effectiveness of machine learning models in the context of classifying SARS-CoV-2 sequencing reads samples. Authors in [19] use probabilistic sequence signatures to get accurate metagenomic reads binning. Their study aims to divide the reads samples into independent sets so that k-mer frequencies are not overestimated.

Designing fixed length feature embeddings is a popular research area in many domains such as graph analytics [1,4], smart grid [2,3], electromyography [50], clinical data analysis [10], network security [6], and text classification [46]. There is another domain of research that deals with the problem of sequence classification, which involves generating a kernel matrix (also called a gram matrix) that contains the distance between pairs of sequences [16,26]. These kernel matrices can be used as input to the kernel based classifiers such as support vector machine (SVM) to perform the sequence classification. Although kernel based methods have been proven to perform well in terms of prediction, one drawback of this approach is the exponential memory cost [9]. Since a kernel matrix is of size $n \times n$ where n is the number of sequences, it becomes almost impossible to store the kernel matrix in memory when n is a large number (e.g., for a million sequences), as shown in [9].

3   Proposed Approach

In this section, we first discuss the approach for generating a consensus sequence from a sequencing reads sample. Then we discuss the different embedding methods we use to generate a feature vector representation. Finally, we discuss the clustering algorithms we used in this paper, including the Pangolin tool as a baseline for comparison.

3.1   Producing Consensus Sequences

We downloaded 6812 raw high-throughput reads samples of COVID-19 patient nasal swab PCR tests from the NCBI SARS-CoV-2 SRA runs resource[1] (see Sect. 4 for some basic descriptive statistics on these data). When then obtained the reference genome from Ensembl[2], and aligned each sample to this reference genome, called the variants in this sample with respect to the reference genome, and then inserted these variants into the reference genome to generate a consensus sequence. The basic steps in more detail are depicted in Fig. 1. We implemented these steps in a Snakemake [32] pipeline, which is available online for reproducibility[3].

3.2   Embedding Approaches

k-mers Based Embedding. In this section, we discuss our alignment-free k-mers based approach, which computes k-mers directly from the reads sample. The k-mers are sub-strings of length k extracted from the reads using a sliding window, as shown in Fig. 2a. The k-mers allow preserving some of the sequential ordering information on the nucleotides within each read. From a read of length N we extract $N - k + 1$ k-mers.

These generated k-mers are then used to create a fixed-length feature vector by taking the frequency of each k-mer. The length of this feature vector is $|\Sigma|^k$

---

[1] https://www.ncbi.nlm.nih.gov/sars-cov-2/.

[2] https://covid-19.ensembl.org/index.html.

[3] https://github.com/murraypatterson/ncbi-sra-runs-pipeline.

Fig. 1. Pipeline for producing consensus sequences. (**1**) We download the reads samples using the NCBI command line batch Entrez tool `fasterq-dump` (https://rnnh.github.io/bioinfo-notebook/docs/fasterq-dump.html), and we download the SARS-CoV-2 reference genome (Wuhan-Hu-1) `GCA_009858895.3.fasta`. (**2**) Each sample is aligned to this reference genome using `bwa mem` [28] and then compressed to BAM (https://samtools.github.io/hts-specs/SAMv1.pdf) format using `samtools` [13]. (**3**) From the sample, an "mpileup" is generated with `bcftools mpileup` [13], variants are then called with `bcftools call`, and the resulting VCF (https://samtools.github.io/hts-specs/VCFv4.2.pdf) file is normalized with `bcftools norm`. (**4**) Finally, the consensus FASTA sequence is generated by inserting into the reference genome the variants from the VCF file generated in the previous step, with `bcftools consensus`.

where Σ is the character alphabet (and k is the length of the k-mers). In our experiments we took $k = 3$ and the alphabet is the nucleotides $\{A, C, G, T\}$. Therefore the feature vector length is $4^3 = 64$.

Minimizers Based Embedding. A minimizer [40] is the lexicographical smallest k-mer in forward and reverse order within a window of size w. In our case, we just compute the minimizer of each short read (the window is the read), as shown in Fig. 2b, allowing for a much more compact frequency vector, as compared to computing all of the k-mers.

These generated minimizers are then used to create a fixed-length feature vector in the same way as in the k-mers based embedding. Clearly, this approach discards almost entirely the sequences, since it preserves only a representative k-mer for each short read.

One-Hot Embedding (OHE) [27]. As a baseline embedding, we use the one-hot encoding (OHE) [27]. This approach generates a binary vector for each nucleotide of $\{A, C, G, T\}$ of the consensus sequence of a sample of reads, where the vector associated with nucleotide N will have 1s for the positions in this consensus sequence that correspond to N, and all other positions will have value 0. Such binary vectors are generated for all nucleotides and are concatenated to form a single vector.

3.3   Clustering Algorithms

Each of the three feature embeddings of the previous section can be input to any of the clustering algorithms that we specify next.

k**-Means** [29]. In order to get the optimal number of clusters for k-means and k-modes, we have used the elbow method [44], which makes use of the knee point detection algorithm (KPDA) [44], as depicted in Fig. 3. From this figure, it is clear that 5 is the optimal number of clusters.

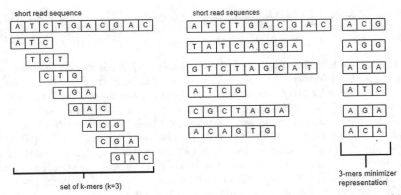

(a) k-mers for read "ATCTGACGAC"(b) Set of minimizers for a reads sample

Fig. 2. Example of k-mers and minimizers.

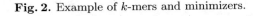

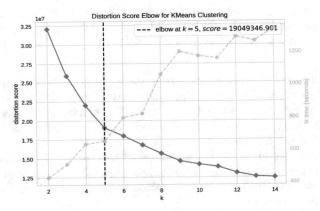

Fig. 3. Use of the elbow method for determining the optimal number of clusters.

k-**Modes** [23]. We also cluster with k-modes, which is an extension of k-means using *modes* instead of *means*. Here, pairs of objects are subject to a dissimilarity measure (*e.g.*, Hamming distance) rather than the Euclidean mean. We have used the same value of k as for k-means.

Pangolin [37]. We also use the Pangolin tool, which takes directly as input the consensus sequence of a sample of reads, as a baseline for comparison. The Pangolin tool assigns the most likely lineage [35] (called the Pango lineage) to a SARS-CoV-2 genome sequence. The Pangolin dynamic nomenclature [38] was devised for identifying SARS-CoV-2 genetic lineages of epidemiological significance and is used by researchers and public health authorities throughout the world to track SARS-CoV-2 transmission and dissemination.

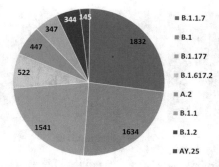

	B.1.1.7
	B.1
	B.1.177
	B.1.617.2
	A.2
	B.1.1
	B.1.2
	AY.25

Fig. 4. Variants distribution for the NCBI reads samples.

Table 1. Variants distribution for a total of 6812 samples, in tabular form.

Variant name	Region	Label	Num. of sequences
B.1.1.7	UK [18]	Alpha	1832
B.1	-	-	1634
B.1.177	-	-	1541
B.1.617.2	India [55]	Delta	522
A.2	-	-	447
B.1.1	-	-	347
B.1.2	-	-	344
AY.25	India [43]	Delta	145

4   Experimental Evaluation

In this section, we first give some details on the data that we use in this study. We then discuss different internal clustering evaluation metrics that we use to measure the performance of the clustering algorithms. Finally, we detail several clustering comparison measures, such as the adjusted Rand index, that we use to compare each pair of clusterings.

4.1   Dataset Statistics and Visualization

The dataset statistics (the labels are assigned to the samples using the Pangolin tool) for our experiments on raw high-throughput reads samples are given in Fig. 4 and Table 1. Both the dataset[4] and the code used in this paper are available online[5].

A first analysis is to check if there is any natural clustering or hidden pattern in the data. However, it is very difficult to visually analyze the information in higher dimensions (*i.e.*, dimensions > 2). For this purpose, we mapped the data to 2-dimensional real vectors using the t-distributed stochastic neighbor embedding (t-SNE) approach [51]. The t-SNE plots for different embedding methods are given in Fig. 5. For OHE (Fig. 5a), we can observe that some data is separated into different clusters, such as in the case of variants 'A2' and 'B.1.1.7'. For the t-SNE plots of k-mers and minimizers (Fig. 5b and 5c respectively), since we are computing the feature embedding directly from the raw reads sample data, we can observe the difference in the structure of data as compared to OHE. Although there is some overlapping between different variants, we can still observe some separation between a few variants, such as 'B.1.1.7' and 'B.1.617.2'. Since there is no clear separation between different variants in any of the t-SNE plots (apart from some small groups), clustering this dataset is not easy.

---

[4] https://drive.google.com/drive/folders/1i4uRrnkjkwUA93EOl8YORBBLb7yIFIm1?usp=sharing.

[5] https://github.com/murraypatterson/ncbi-sra-runs-pipeline.

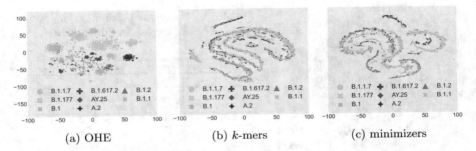

(a) OHE   (b) k-mers   (c) minimizers

Fig. 5. t-SNE plots for different embedding methods (labels by Pangolin). This figure is best seen in color.

4.2   Evaluation Metrics

In this section, we describe the internal clustering evaluation measures used to assess the quality of clustering.

Silhouette Coefficient [42]. Given a feature vector, the silhouette coefficient computes how similar the feature vector is to its own cluster (cohesion) compared to other clusters (separation) [36]. Its score ranges between $[-1, 1]$, where 1 means the best possible clustering and -1 means the worst possible clustering.

Calinski-Harabasz Score [12]. The Calinski-Harabasz score evaluates the validity of a clustering based on the within-cluster and between-clusters dispersion of each object with respect to each cluster (based on sum of squared distances) [36]. A higher score denotes better defined clusters.

Davies-Bouldin Score [14]. Given a feature vector, the Davies-Bouldin score computes the ratio of within-cluster to between-cluster distances [36]. A smaller score denotes groups are well separated, and the clustering results are better.

4.3   Clustering Comparison Metrics

To compare different clustering approaches, we use the following metrics:

Adjusted Rand Index [24]. Given two clusterings, the adjusted Rand index (ARI) computes the similarity between them by considering all pairs of clusters output and counts the pairs that are assigned to the same or different clusters. The value ranges between $[0, 1]$, where 1 denotes an identical labeling, and approaches 0 as they become more different. A pair of random labelings has an expected ARI of almost 0.

Fowlkes-Mallows Index [17]. Given two clusterings, the Fowlkes-Mallows index (FMI) first computes the confusion matrix for the clustering output. The FMI is then defined by the geometric mean of the precision and the recall. A larger value of the FMI indicates a greater similarity between the clusters.

Completeness Score [41]. Given two clustering outputs, the completeness score (CS) evaluates how much similar samples are placed in the same cluster. Its value ranges between $[0, 1]$, where 1 means complete clustering agreement and approaches 0 the further it deviates from this.

V-Measure [41]. Given two clustering outputs, we first compute homogeneity (evaluate if objects belong to the same or different cluster) and completeness (evaluate how much similar samples are placed together by the clustering algorithm). The V-measure is then defined by the harmonic mean of homogeneity and completeness. This score is a number between $[0, 1]$ where 1 indicates a perfect clustering and approaches 0 the further it deviates from this.

5   Results and Discussion

In this section, we first report the results for all combinations of clustering and embedding method of Sect. 3, using the internal clustering metrics of Sect. 4. We then compare all clusterings using the comparison metrics of Sect. 4. We then measure, in terms of information gain, the importance of each nucleotide position to the label assigned by the combination above which had the best overall internal clustering score. Finally, we perform a statistical analysis using Pearson and Spearman correlation to evaluate the compactness of each feature embedding.

5.1   Clustering Evaluation

Table 2 shows the results for different clustering algorithms and their comparison with various embeddings on three internal clustering evaluation metrics. Since Pangolin [37] takes sequences as input rather than numerical feature vectors, we cluster the sequences using the Pangolin tool and then we evaluate the quality of the clustering labels using the different numerical embeddings of these sequences. The performance of OHE with the Pangolin labels is better compared to the performance of k-mers or minimizers with Pangolin labels overall, probably because OHE is a straightforward numerical representation of a sequence (possibly similar to the ML representation that Pangolin uses internally). Moreover, we can observe that the k-mers based feature embedding performs better with k-means clustering in all but one evaluation metric. An important observation here is that the clustering from k-means shows better performance as compared to the Pangolin tool overall. This indicates that the Pangolin tool may not be the best option in this raw high-throughput sequencing reads setting. We observe that the k-modes clustering algorithm performs poorly in most cases for clustering and runtime.

Table 2. Internal clustering quality metrics for Pangolin, k-means and k-modes on OHE, k-mers and minimizers embeddings. Best values are shown in bold.

Algorithm	Embedding	Evaluation metrics			Clustering runtime
		Silhouette coefficient	Calinski-Harabasz score	Davies-Bouldin score	
Pangolin [37]	OHE [27]	0.029	818.673	8.471	≈14 h
	k-mers	−0.214	78.098	7.864	
	minimizers	−0.233	144.200	6.063	
k-means	OHE [27]	0.623	3278.376	1.502	648.5 s
	k-mers	0.775	**21071.221**	**0.406**	**19.2 s**
	minimizers	**0.858**	17909.284	0.421	31.3 s
k-modes	OHE [27]	0.613	2053.754	2.056	≈ 7 days
	k-mers	−0.027	9.801	89.789	≈4 h
	minimizers	−0.398	1196.777	3.545	≈1 h

5.2 Comparing Different Clusterings

We compare the different clustering algorithms on different embeddings using the adjusted Rand index (ARI), Fowlkes-Mallows index (FMI), V-measure (VM) and completeness score (CS). The heat map in Fig. 6 shows that some embeddings are more similar to others. We observe that the k-mers + k-means combination is very similar to the minimizers + k-means combination in terms of ARI, FMI, VM and CS. This shows that the clusterings from the combinations of clustering method and embedding are not much different from each other.

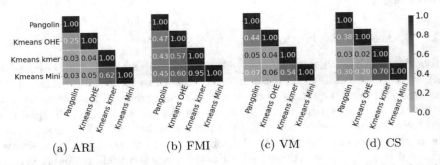

Fig. 6. Comparison of different clustering approaches and embedding methods using standard clustering comparison metrics.

5.3 Information Gain

The combination of clustering method and embedding which performed the best overall, in terms of internal clustering quality, was k-mers + k-means (see Table 2). To verify if such a clustering makes sense from a biological standpoint, thereby independently validating such a clustering from an orthogonal viewpoint, we also computed the importance of each genomic position in the sequence to the labeling (the clustering) obtained by k-mers + k-means. For this purpose, we computed the *information gain* (IG) of this labeling in terms of genomic position, defined as:

$$IG(\text{class}, \text{position}) = H(\text{class}) - H(\text{class} \mid \text{position}) \tag{1}$$

where

$$H(\text{class}) = \sum_{e \in \text{class}} -p_e \log p_e \tag{2}$$

is the entropy of a class in terms of the proportion of each unique label e of this class. Figure 7 shows the IG values for different genomic positions corresponding to the class labels. We can see that many positions have higher IG values, which means that they play an important role in predicting the labels.

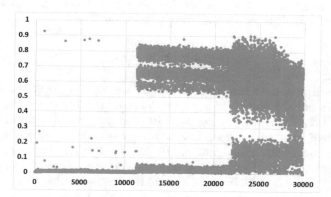

Fig. 7. Information gain values for different genomic positions. The x-axis shows the position and the y-axis shows the information gain value.

This IG scatter plot partitions the SARS-CoV-2 genome into three distinct regions, with very low IG (0–11 Kbp), with either very high or very low IG (11 Kbp–22 Kbp), and with a wide range of IG (22 Kbp–30 Kbp). What is interesting is that the structural proteins S, E, M and N [52] fall in the 21 Kbp–25 Kbp range, overlapping, for the most part, this third region with the wide range of IG. This is consistent with the observation that mutations of the SARS-CoV-2 genome (which define many of the different variants) appear disproportionately in the structural proteins region, particularly the spike (S) region [54].

5.4 Statistical Analysis

Since the information gain is in terms of positions of the genomic sequence, it does not provide information on how important the *features* of each embedding are (since feature vectors are in the Euclidean space). For this purpose, we use Spearman and Pearson correlation to evaluate the (negative and positive) importance of features in the different embeddings. Since we note from the previous Sect. 5.2 that the combination of k-mers + k-means is quite similar to minimizers + k-means, we performed such an analysis on the consensus (agreement of both clusterings) labels from both clusterings. A total of 5738 labels and the corresponding feature embedding were analyzed.

Pearson Correlation. We also use Pearson correlation [11] to evaluate the compactness of different feature embeddings. The Pearson correlations are shown in Fig. 8a for the negative correlation range and Fig. 9a for the positive correlation range. For negative correlation, we can observe that OHE has more features corresponding to the consensus labeling, with a higher Pearson correlation. However, in the case of positive correlation, the k-mers based feature vector seems to be more compact than OHE and minimizers based feature embedding.

Spearman Correlation. We use Spearman correlation [33] to evaluate the contribution of different attributes of feature embeddings. The Spearman Correlation is computed using the following expression:

$$\rho = 1 - \frac{6 \sum d_i^2}{n(n^2 - 1)},\tag{3}$$

where ρ is the Spearman's rank correlation coefficient, d_i is the difference between the two ranks of each observation, and n is the total number of observations.

The number of features having negative and positive Spearman correlation for the different ranges for different embedding methods are shown in Fig. 8b for negative the correlation range and Fig. 9b for the positive correlation range. In terms of negative correlation, we can observe that there is only a small fraction of features in the case of OHE. At the same time, other embeddings do not have any features which are negatively correlated to the consensus labeling. In the case of positive correlation, we can observe that the k-mers based embedding has more features with high correlation values than the other embeddings. This indicates that the feature embedding from k-mers is more compact than OHE and the minimizers based feature embedding approach.

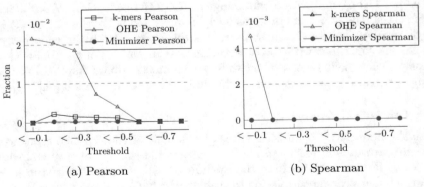

(a) Pearson   (b) Spearman

Fig. 8. Fraction of features negatively correlated with the labeling in the case of Pearson and Spearman. This figure is best seen in color.

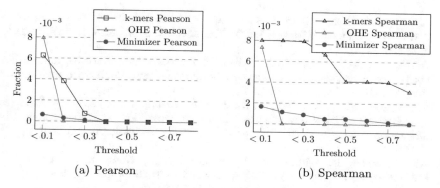

Fig. 9. Fraction of features positively correlated with the labeling in the case of Pearson and Spearman. This figure is best seen in color.

6   Conclusion

We propose the use of alignment-free feature vector embedding approaches in the setting of raw high-throughput reads data. Such embeddings are then used as input to different clustering methods such as k-means and k-modes. Using internal clustering evaluation metrics, we indicate that the proposed embeddings are more suited to this sequencing reads setting than the widely accepted Pangolin tool. We also show that the proposed feature embeddings are better in terms of runtime than more traditional embedding methods such as OHE, on top of the fact that they avoid the expensive genome assembly step. Finally, in computing the information gain (IG), we show that most of the genomic positions having high importance (high IG) corresponding to the labeling (from the clustering) are concentrated in the spike region of the SARS-CoV-2 genome.

In the future, we would explore the scalability of such approaches by using more data. Another direction of future work is to explore how sensitive the predictions of the Pangolin tool, or OHE are to the genome assembly. Finally, we would like to explore the usage of such embedding in conjunction with alignment-free variant calling, which could possibly eliminate even more dependencies on the genome assembly step.

References

1. Ahmad, M., Ali, S., Tariq, J., et al.: Combinatorial trace method for network immunization. Inf. Sci. **519**, 215–228 (2020)
2. Ali, S., Mansoor, H., Khan, I., Arshad, N., Khan, M.A., Faizullah, S.: Short-term load forecasting using AMI data. arXiv:1912.12479 (2019)
3. Ali, S., Mansoor, H., et al.: Short term load forecasting using smart meter data. In: International Conference on Future Energy Systems (e-Energy), pp. 419–421 (2019)

4. Ali, S., Shakeel, M., Khan, I., Faizullah, S., Khan, M.: Predicting attributes of nodes using network structure. ACM Trans. Intell. Syst. Technol. (TIST) **12**(2), 1–23 (2021)
5. Ali, S., Ali, T.E., Khan, M.A., Khan, I., Patterson, M.: Effective and scalable clustering of SARS-CoV-2 sequences. In: International Conference on Big Data Research (ICBDR), pp. 42–49 (2021)
6. Ali, S., Alvi, M.K., Faizullah, S., Khan, M.A., Alshanqiti, A., Khan, I.: Detecting DDoS attack on SDN due to vulnerabilities in OpenFlow. In: International Conference on Advances in the Emerging Computing Technologies (AECT), pp. 1–6 (2020)
7. Ali, S., Bello, B., Chourasia, P., Punathil, R.T., Zhou, Y., Patterson, M.: PWM2Vec: an efficient embedding approach for viral host specification from coronavirus spike sequences. Biology **11**(3), 418 (2022)
8. Ali, S., Patterson, M.: Spike2Vec: an efficient and scalable embedding approach for Covid-19 spike sequences. In: IEEE International Conference on Big Data (Big Data), pp. 1533–1540 (2021)
9. Ali, S., Sahoo, B., Ullah, N., Zelikovskiy, A., Patterson, M., Khan, I.: A k-mer based approach for SARS-CoV-2 variant identification. In: International Symposium on Bioinformatics Research and Applications, pp. 153–164 (2021)
10. Ali, S., Zhou, Y., Patterson, M.: Efficient analysis of Covid-19 clinical data using machine learning models. Med. Biol. Eng. Comput. 1–16 (2022)
11. Benesty, J., Chen, J., Huang, Y., Cohen, I.: Pearson correlation coefficient. In: Benesty, J., Chen, J., Huang, Y., Cohen, I. (eds.) Noise Reduction in Speech Processing. Springer Topics in Signal Processing, vol. 2, pp. 1–4. Springer, Heidelberg (2009). https://doi.org/10.1007/978-3-642-00296-0_5
12. Calinski, T., Harabasz, J.: A dendrite method for cluster analysis. Commun. Stat.-Theory Methods **3**(1), 1–27 (1974)
13. Danecek, P., et al.: Twelve years of SAMtools and BCFtools. GigaScience **10**(2) (2021)
14. Davies, D., Bouldin, D.: A cluster separation measure. IEEE Trans. Pattern Anal. Mach. Intell. **2**, 224–227 (1979)
15. Du Plessis, L., et al.: Establishment and lineage dynamics of the SARS-CoV-2 epidemic in the UK. Science **371**(6530), 708–712 (2021)
16. Farhan, M., Tariq, J., Zaman, A., Shabbir, M., Khan, I.: Efficient approx algorithms for strings kernel based sequence classification. In: Advances in Neural info Processing System (NeurIPS), pp. 6935–6945 (2017)
17. Fowlkes, E., Mallows, C.: A method for comparing two hierarchical clusterings. J. Am. Statist. Assoc. **78**(383), 553–569 (1983)
18. Galloway, S., et al.: Emerg. of SARS-CoV-2 b.1.1.7 lin. Morb. Mortal. Weekly Repo. **70**(3), 95 (2021)
19. Girotto, S., Pizzi, C., Comin, M.: MetaProb: accurate metagenomic reads binning based on probabilistic sequence signatures. Bioinformatics **32**(17), i567–i575 (2016)
20. GISAID. https://www.gisaid.org/. Accessed 29 Jan 2022
21. Golubchik, T., Wise, M., Easteal, S., Jermiin, L.: Mind the gaps: evidence of bias in estimates of multiple sequence alignments. Mol. Biol. Evol. **24**(11), 2433–2442 (2007)
22. Hadfield, J., et al.: Nextstrain: real-time tracking of pathogen evolution. Bioinformatics **34**, 4121–4123 (2018)
23. Huang, Z.: Extensions to the k-modes algorithm for clustering large data sets with categorical values. Data Min. Knowl. Disc. **2**(3), 283–304 (1998)

24. Hubert, L., Arabie, P.: Comparing partitions. J. Classification **2**(1), 193–218 (1985)
25. Kawulok, J., Deorowicz, S.: CoMeta: classification of metagenomes using k-mers. Plos One (2015)
26. Kuksa, P., Khan, I., Pavlovic, V.: Generalized similarity kernels for efficient sequence classification. In: SIAM International Conference on Data Mining (SDM), pp. 873–882 (2012)
27. Kuzmin, K., et al.: Machine learning methods accurately predict host specificity of coronaviruses based on spike sequences alone. Biochem. Biophys. Res. Commun. **533**(3), 553–558 (2020)
28. Li, H.: Aligning sequence reads, clone sequences and assembly contigs with BWA-MEM (2013)
29. Lloyd, S.: Least squares quantization in PCM. IEEE Trans. Inf. Theory **28**(2), 129–137 (1982)
30. Melnyk, A., et al.: From alpha to zeta: identifying variants and subtypes of SARS-CoV-2 via clustering. J. Comput. Biol. **28**(11), 1113–1129 (2021)
31. Minh, B., et al.: IQ-TREE 2: new models and efficient methods for phylogenetic inference in the genomic era. Mol. Biol. Evol. **37**, 1530–1534 (2020)
32. Mölder, F., Jablonski, K.P., Letcher, B., et al.: Sustainable data analysis with Snakemake. F1000Res **10**(33) (2021)
33. Myers, L., Sirois, M.: Spearman correlation coefficients, differences between. Encyclopedia Stat. Sci. **12** (2004)
34. Needham, K.: Chinese state fund invests in gene firm BGI. Reuters [Internet] (2021). https://www.reuters.com/article/us-china-genomics-state-idUSKBN2AM0AT
35. O'Toole, A., et al.: Assignment of epidemiological lineages in an emerging pandemic using the pangolin tool. Virus Evol. **7**(2), veab064 (2021)
36. Pedregosa, F., et al.: Scikit-learn: machine learning in Python. J. Mach. Learn. Res. **12**, 2825–2830 (2011)
37. Phylogenetic Assignment of Named Global Outbreak Lineages (Pangolin). https://cov-lineages.org/resources/pangolin.html. Accessed 4 Jan 2022
38. Rambaut, A., et al.: A dynamic nomenclature proposal for SARS-CoV-2 lineages to assist genomic epidemiology. Nat. Microbiol. **5**(11), 1403–1407 (2020)
39. Reporter, S.: CDC commits $90m to create public health pathogen genomics research centers. Genomeweb. https://www.genomeweb.com/infectious-disease/cdc-commits-90m-create-public-health-pathogen-genomics-research-centers. Accessed 29 Jan 2022
40. Roberts, M., Hayes, W., Hunt, B., Mount, S., Yorke, J.: Reducing storage req for biological sequence comparison. Bioinformatics **20**(18), 3363–3369 (2004)
41. Rosenberg, A., Hirschberg, J.: V-measure: a conditional entropy-based external cluster evaluation measure. In: The Joint Conference Empirical Methods NLP Computational Natural Language Learning (EMNLP-CoNLL), pp. 410–420 (2007)
42. Rousseeuw, P.: Silhouettes: a graphical aid to interpretation and validation of cluster analysis. J. Comput. Appl. Math. **20**, 53–65 (1987)
43. SARS-CoV-2 Variant Classifications and Definitions. https://www.cdc.gov/coronavirus/2019-ncov/variants/variant-info.html. Accessed 29 Jan 2022
44. Satopaa, V., Albrecht, J., Irwin, D., Raghavan, B.: Finding a "Kneedle" in a haystack: detecting knee points in system behavior. In: International Conference on Distributed Computing Systems Workshops, pp. 166–171 (2011)
45. Sboner, A., Mu, X., Greenbaum, D., Auerbach, R., Gerstein, M.: The real cost of sequencing: higher than you think! Genome Biol. **12**(8), 125 (2011)

46. Shakeel, M., Faizullah, S., Alghamidi, T., Khan, I.: Language independent senti-
    ment analysis. In: International Conference on Advances in the Emerging Com-
    puting Technologies (AECT), pp. 1–5 (2020)
47. Solis, S., Avino, M., Poon, A., Kari, L.: An open-source k-mer based machine
    learning tool for fast and accurate subtyping of HIV-1 genomes (2018)
48. Stephens, Z., et al.: Big data: astronomical or genomical? PLoS Biol. **13**(7),
    e1002195 (2015)
49. Tayebi, Z., Ali, S., Patterson, M.: Robust representation and efficient feature selec-
    tion allows for effective clustering of SARS-CoV-2 variants. Algorithms **14**(12), 348
    (2021)
50. Ullah, A., Ali, S., Khan, I., Khan, M.A., Faizullah, S.: Effect of analysis window and
    feature selection on classification of hand movements using EMG signal. In: Arai,
    K., Kapoor, S., Bhatia, R. (eds.) IntelliSys 2020. AISC, vol. 1252, pp. 400–415.
    Springer, Cham (2021). https://doi.org/10.1007/978-3-030-55190-2_30
51. Van der Maaten, L., Hinton, G.: Visualizing data using t-SNE. J. Mach. Learn.
    Res. (JMLR) **9**(11) (2008)
52. Walls, A., Park, Y., Tortorici, M.: Structure, function and antigenicity of the SARS-
    CoV-2 spike glycoprotein. Cell **181**(2), 281–292 (2020)
53. Wood, D., Salzberg, S.: Kraken: ultrafast metagenomic sequence classification
    using exact alignments. Genome Biol. **15** (2014)
54. Xu, W., Wang, M., Yu, D., Zhang, X.: Variations in SARS-CoV-2 spike protein cell
    epitopes and glycosylation profiles during global transmission course of Covid-19.
    Front. Immunol. **11** (2020)
55. Yadav, P., et al.: Neutralization potential of Covishield vaccinated individuals sera
    against B.1.617.1. Clin. Infect. Dis. **74**, 558–559 (2021)

Analysis of SARS-CoV-2 Temporal Molecular Networks Using Global and Local Topological Characteristics

Fiona Senchyna and Rahul Singh[⌧]

Department of Computer Science,
San Francisco State University, 1600 Holloway Avenue, San Francisco, CA 94132, USA
fsenchyna@mail.sfsu.edu, rahul@sfsu.edu

Abstract. The global COVID-19 pandemic continues to have a devastating impact on human population health. In an effort to fully characterize the virus, a significant volume of SARS-CoV-2 genomes have been collected from infected individuals and sequenced. Comprehensive application of this molecular data toward epidemiological analysis in large parts has employed methods arising from phylogenetics. While undeniably valuable, phylogenetic methods have their limitations. For instance, due to their rooted structure, outgroup samples are often needed to contextualize genetic relationships inferred by branching. In this paper we describe an alternative: global and local topological characterization of neighborhood graphs relating viral genomes collected from samples in longitudinal studies. The applicability of our approach is demonstrated by constructing and analyzing such graphs using two distinct datasets from Israel and France, respectively.

Keywords: SARS-CoV-2 · Graph topology · Network theory · Computational epidemiology

1 Introduction

The rapid dissemination of Coronavirus-Diesase-19 (COVID-19) since its first recorded outbreak in December 2019 has led to a worldwide pandemic with devastating consequences. According to the World Health Organization (WHO), currently over 300 million cases have been confirmed globally, including over 5 million deaths [1]. Consequently, the global research community has taken tremendous efforts to study the etiological agent, SARS-CoV-2, of COVID-19. The unprecedented volume of whole genomes sequenced and made publicly available has led to rapid advancements in areas such as drug development, diagnostics, and understanding of the pathogenicity and epidemiology of the virus [2–6]. To date, the GISAID (Global Initiative on Sharing Avian Influenza Data) database, a popular publicly available repository for SARS-CoV-2 sequence data, contains over 11 million genomes [7].

In molecular epidemiology, genomes sampled from infected individuals are related to one another based on sequence similarity, typically within a phylogenetic framework.

© The Author(s), under exclusive license to Springer Nature Switzerland AG 2022
M. S. Bansal et al. (Eds.): ICCABS 2021, LNBI 13254, pp. 149–162, 2022.
https://doi.org/10.1007/978-3-031-17531-2_12

Inferences are then made regarding the spread and prevalence of a virus within a population. Although originally intended to determine the relatedness of different taxa, phylogenetics has been co-opted and modified for analysis of pathogen transmission [8–10]. Phylodynamic analysis, which studies the interaction and influence of epidemiological, immunological, and evolutionary processes on viral evolution and genetic variation, similarly infers viral population levels over time based on a phylogenetic tree [11]. Regarding SARS-CoV-2, phylogenetic and phylodynamic studies have been used to estimate the source and date of origin of infection, the temporal reproductive number, geographical spread, and the role of super spreaders [5, 6, 9].

Although demonstrably valuable, phylogenetics has its limitations when analyzing disease spread. Phylogenetics methods, particularly those employing Bayesian models, have many parameters that can be challenging to estimate (*e.g.*, the substitution model, molecular clock, and priors). Often, the phylogenetic tree has to be re-computed if a new sequence is added [12] – a significant overhead in large epidemiological settings. Additionally, the constraint of a tree structure limits the topology of the patterns that can be hypothesized and studies. Indeed, the topology of infection spread generally does not conform to the constraints of a single source and predefined branching tree structure [13]. By contrast, a different view of the information arises when it is modeled as a network (graph). Network representations of data have allowed for increased understanding of several biological phenomena (*e.g.*, gene and protein functions, human neural networks, and epidemiological contact tracing) [13–17]. Network properties derived from such representations can be divided into "global" and "local". Global network properties include degree distribution, diameter, path length, and centrality and characterize the connectivity of the entire dataset. Local properties on the other hand, characterize a network in terms of the connectivity of its node to nodes in a local neighborhood. Small, induced subgraphs, called *graphlets*, of the larger network, are one such local topological property. In practice, graphlets are typically defined to consist of graphs containing 3 to 5 nodes. This yields 29 unique graphlet structures whose presence can be used to characterize the local structure of a network. The *relative graphlet frequency* (RGF) distance between two networks can be used as a network comparison measure by comparing the frequency of all 29 *graphlets* in both networks [14]. Local connectivity can also be investigated with the graph Laplacian, which partitions the graph based on an optimal cut, and can reveal communities of nodes within the graph [18].

Molecular genetic networks have been computed for Human Immunodeficiency virus (HIV) and Hepatitis C virus (HCV) for contact tracing purposes [16, 19–21]. In these networks, viral genetic samples taken from infected individuals are represented as nodes. Edges are added between two nodes if the genetic distance between the pair of samples is below a certain distance threshold. HIV and HCV are blood-borne viruses, and their transmission is often associated with high-risk behaviors [22, 23]. In contrast, SARS-CoV-2 is a highly transmissible airborne virus. The resulting large volume of unsampled hosts makes it virtually impossible to accurately perform contact tracing from sampled sequences alone [23]. However, analysis of changes in topological properties of a SARS-CoV-2 genetic network can provide insight about the accumulation (or lack thereof) in variation of the virus within a population over time. Here, temporal genetic networks were built for two datasets separately based on a genetic distance threshold of

2×10^{-4}. Global and local properties of the graphs were analyzed to characterize the dataset and relate to underlying biological changes, including the use of graph cuts to identify emerging viral subtypes within the datasets.

2 Data and Methods

2.1 Data and Preprocessing

Analysis of SARS-CoV-2 molecular evolution within a population was performed on two distinct datasets described previously [5, 6]. Samples in each dataset originated from France and Israel, respectively, and were collected during the first wave of the pandemic in the early months of 2020. For each sequence, the collection date was known. SARS-CoV-2 genomes were downloaded from the GISAID database (https://www.gisaid.org). Accession numbers for the Israel dataset (IDS) ($n = 212$) are EPI_ISL_447258 - EPI_ISL_447469; and for French dataset (FDS) ($n = 186$) are EPI_ISL_414624-7,29-38, EPI_ISL_415649-54, EPI_ISL_416493-502,504-506, 508-513, EPI_ISL_416745-52, 54, 56-58, EPI_ISL_417333-4, 36-40, EPI_ISL_418218-40, EPI_ISL_418412-31, EPI_ISL_419168-88, EPI_ISL_420038-64, EPI_ISL_420604-25, and EPI_ISL_421500-1. A reference sequence (originating from the first recorded outbreak in Wuhan, China) was downloaded from GenBank (https://www.ncbi.nlm.nih.gov/genbank/, accession number MN908947). Separately, the genomes for each dataset were aligned to the reference sequence with MAFFT [24]. Non-coding regions were removed from all genomes according to the reference sequence annotation. Additionally, samples were removed from further downstream analysis if the coding region contained more than 1% ambiguous nucleotides or gaps. Insertions and deletions (indels) were ignored due to lack of clarity between indels and ambiguous nucleotides. The final size consisted of 171 and 181 samples for IDS and FDS, respectively. After removal of non-coding regions, sequences had a nucleotide length of 29,132.

2.2 Construction of Temporal Networks

Each dataset was represented as a set of temporally evolving networks,

$$G(t) = (V(t), E(t)) \tag{1}$$

where $t = \{t_1, \dots t_n\}$ are the set of time points corresponding to the sample collection times. Consequently, $\Delta G = G(t_i) - G(t_{i-1})$ defines the incremental change in the network between the times t_i and t_j and $V(t) = \{v_1, \dots, v_n\}$ represents the samples collected on or before t. The edge, $e_{ij} \in E(t)$ connects v_i and $v_j \in V(t)$ if the genetic distance between v_i and v_j is below a threshold and indicates the respective samples to be genetically close within the time spanned by t. The reader may note that no constraints are placed on the specific time-points at which the data is gathered.

Pairwise genetic distances are calculated as the number of sites where the two sequences differed divided by the total length of the sequence (hamming distance). For each pairwise comparison, positions with ambiguous nucleotides or gaps are ignored. A genetic distance threshold of 2×10^{-4} was empirically chosen to connect sample nodes in the network. This threshold ensured that the majority of pairwise distances were below the cutoff value in both datasets: 53.8% (7,813/14,535) for IDS and 70.8% (11,532/16,290) for FDS (Fig. 1). Nevertheless, the threshold was high enough to prevent formation of highly connected networks that lacked any meaningful topology. IDS produced 27 graphs from March 17th to April 22nd, 2020. FDS produced 26 graphs dated from February 26th to March 24th.

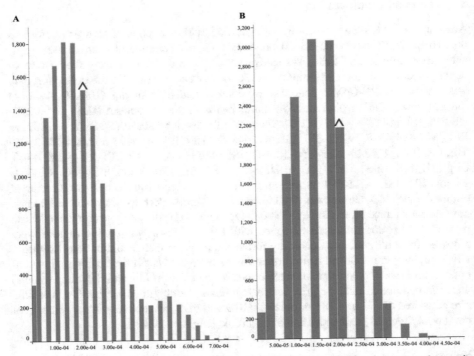

Fig. 1. Genetic distance distribution for (A) IDS and (B) FDS. Genetic distances are on the x-axis and number of pairwise comparisons are on the y-axis. ^ indicates the bin containing the cut-off threshold in both plots.

2.3   Global Network Analysis

To assess how nucleotide changes in the viral population are reflected in a network, the following global properties were calculated: degree distribution, average clustering coefficient, average path length, and diameter. Centrality measures including degree, closeness, and betweenness centrality were also calculated to identify those nodes most central to the network and relate to the nucleotide constitution of the respective datasets.

2.4   Local Topological Analysis

To ensure that small inconsequential connected components do not impact comparisons of networks, only the largest subnetwork, $s_i(G(t))$, for $G = (G_1, \ldots, G_n)$ was used for local connectivity analysis. This was done as the largest subnetwork contained over 98% of total samples in the majority of graphs. Henceforth, the major subnetworks will be referred to as $G_1, \ldots, G_n$. That is, for notational simplicity, we are simply using the notation G_i to denote the largest subnetwork of G_i. We employed the notion of graphlets to conduct the local connectivity analysis using *Graph Crunch 2* [14]. To identify topological changes between sampling periods, the *relative graphlet frequency distance* (RGF distance) was calculated between consecutive graphs, $G(t_i)$ and $G(t_{i+1})$. The RGF distance (D) is defined as follows [25]:

$$D(\Delta, \Omega) = \sum_{i=1}^{29} |F_i(\Delta) - F_i(\Omega)| \qquad (2)$$

where Δ and Ω represent $G(t_i)$ and $G(t_{i+1})$, and $i \in \{1, \ldots, 29\}$ in (2) are the number of distinct *graphlets*. Further,

$$F_i(\Phi) = -\log(N_i(\Phi)) \Big/ \sum_{i=1}^{29} N_i(\Phi) \qquad (3)$$

The function F_i divides the log of the frequency of a graphlet by the sum of the frequencies of all graphlets to ensure the difference in node size between Δ and Ω is accounted for (3).

2.5 Quantification of Nucleotide Variation

The consensus sequences for each $G = (G_1, \ldots, G_n)$ were calculated. Single nucleotide variants (SNVs) were identified for each sample by pairwise comparison to the reference sequence. For any single SNV, if its frequency within the population was above 0.1 it was deemed a mutation of interest. To quantify change over time, nucleotide diversity of G_t and the difference in diversity between G_t and samples added at G_{t+1} (Δ_{t+1}) were calculated. Nucleotide diversity was characterized according to the definition by Nei and Li [26].

2.6 Spectral Network Partitioning

The connected components in each of $G_1, \ldots, G_n$ were split using spectral partitioning. Let the Laplacian matrix of a network G, $L(G)$, be defined as follows:

$$L(G) = D(G) - A(G) \qquad (4)$$

where $A(G)$ is the adjacency matrix and $D(G)$ is diagonal matrix of the graph (4). Partitioning of the graph into connected components was accomplished through eigendecomposition of $L(G)$. Nodes (samples) are split based on whether their values in the eigenvector associated with the second smallest eigenvalue is above or below a defined threshold. Five thresholds were tested for the initial splitting of the graph (-0.008, -0.0075, -0.007, -0.005, and 0). The quality of a partition was quantified using the normalized cut value induced by that partition. A threshold of -0.007 was chosen as it consistently gave the lowest normalized cut value for all graphs in both datasets. The connected components were recursively partitioned into connected components of at least two nodes to investigate further groupings of samples when applicable.

2.7 Phylogenetic Analysis

As a comparison to the spectral partitioning of the graph into clusters, phylogenetic analysis was performed on the genome sequences. A maximum likelihood phylogenetic tree was constructed using RAxML (Randomized Accelerated Maximum Likelihood) v1.0.0 [27] with a GTR substitution model and 100 bootstrap replicates.

3 Results and Analysis

3.1 Genetic Characterization of the Viral Population

IDS. The initial consensus sequence had 4 SNVs compared to the reference sequence. These were C3037T, C14408T, A23403G, and G25563T. On March 21st, the nucleotide at position 1059 in the consensus changed from C to T. The proportion of samples containing a C at this position decreased from 60% (6/10) to 42.86% (6/14) (-28.57%). There were 11 mutations of interest that were not part of the consensus sequence. These included C2416T (overall frequency, 0.11), C11916T (0.17), C18998T (0.15), G28881A (0.11), G28882A (0.11), and G28883C (0.12). The nucleotide diversity between consecutive graphs remained consistent (median, 1.99×10^{-4}, , interquartile range (IQR), 1.83×10^{-4}–2.06×10^{-4}). Similarly, the absolute difference in diversity between samples within G_t and Δ_{t+1} was also very small (median, 5.15×10^{-5}, IQR, 2.41×10^{-5}–7.21×10^{-5}).

FDS. Compared to the reference sequence, the consensus sequence had 4 SNVs (C3037T, C14408T, A23403G, and G25563T). The nucleotide at position 1059 of the consensus changed from T to C on March 4th, the opposite of IDS. The frequency of C in this position increased from 26.67% (4/15) to 51.61% (16/31) (+93.51%). This nucleotide is in the open reading frame (ORF) 1a region of the SARS-CoV-2 genome and encodes the gene *Nsp2*. The reference sequence contains C and the mutation to T is non-synonymous. However, the full functionality of *Nsp2* has yet to be fully understood and so the effect of this mutation on viral fitness is unknown [28, 29].Two SNVs, C2416T (0.13) and C15324T (0.34), were deemed of interest. Like IDS, there was little variation in nucleotide diversity between consecutive graphs (median, 1.28×10^{-4}, IQR, $1.22 \times 10^{-4} - 1.47 \times 10^{-4}$). The absolute difference in diversity between G_t and Δ_{t+1} was also minimal (median, 2.46×10^{-5}, IQR, 1.27×10^{-5}–4.51×10^{-5}).

3.2 Changes in Global Properties

IDS. We found that most global network properties experienced little change. The diameter gradually increased over time, shifting from 2 to 3 then growing to 5 as new samples were added. The median clustering coefficient was 0.88 (IQR, 0.86–0.89) and the median average path length was 1.58 (IQR, 1.55–1.61). There was some change in the most central nodes according to degree and closeness centrality. However, the most central nodes were not vastly different in their genomic constitution. They differed from the consensus sequence by 0–3 nucleotides (EPI_ISL_447408, EPI_ISL_447310,

EPI_ISL_447305, EPI_ISL_447277, EPI_ISL_447284). The joining of a smaller subnetwork to the major subnetwork on March 30th did not influence degree or closeness centrality, but the node with the highest betweenness centrality did shift from EPI_ISL_447277 to EPI_ISL_447447. EPI_ISL_447277 had 3 nucleotide differences from the consensus sequence, while EPI_ISL_447447 had 6. This change is most likely due to the difference in the measures of centrality. Both the degree and closeness centrality measure the relation of a node to all other nodes in the graph, by node degree or length of shortest paths, respectively. Whereas betweenness centrality measures the impact of a node on the shortest paths between all other pairs of nodes in the graph. A divergence in a small proportion of the samples from the consensus sequence would, therefore, naturally have a larger effect on betweenness than closeness or degree. There were no significant findings in the changes in the degree distribution.

FDS. There was little change in the FDS diameter (1–4), average clustering coefficient (median, 0.91, IQR, 0.89–0.92) and average path length (median, 1.17, IQR, 1.15–1.28) over time. Initially, between February 26th and March 2nd, the graph was fully connected and so all samples were equally central. As the graph progressed through time, the nodes with the highest degree, closeness, and betweenness centrality overlapped substantially. These samples differed by 0–3 mutations from the consensus sequence. By the end of the period, the nodes with the highest betweenness centrality were identical to the consensus sequence except at position 25563 (EPI_ISL_414631, EPI_ISL_416494, EPI_ISL_420047). This mutation was present in the population at a frequency of 0.42. Four additional nodes shared the highest degree and closeness centrality. These nodes were an exact match to the consensus sequence (EPI_ISL_418219, EPI_ISL_417336, EPI_ISL_418425, EPI_ISL_419168). Again, the degree distribution did not provide significant insight.

3.3 Association Between RGF Distance and Genetic Variation

The RGF distance between consecutive IDS graphs ranged from 0 to 5.57. Most distances were below 1 (77%, 20/26). However, there were two periods where the RGF distance sharply increased. The first was between March 20th and March 25th, reaching a maximum distance of 5.57, and the second between April 19th and 20th, with a distance of 2.01 (Fig. 2). For FDS, the RGF distance between consecutive graphs ranged from 0.02 to 8.58. Similar to IDS, most distances were below 1 (88%, 22/25). Again, there were two peaks where large distances were recorded: that being 8.58 between March 2nd and 4th, and 1.98 between March 21st and 22nd (Fig. 2). Graphs pertaining to the first peak in RGF distances for IDS and FDS are illustrated in Figs. 3 and 4, respectively.

Interestingly, change in the consensus sequence (described in Sect. 3.1) corresponded with the highest RGF distance for both IDS and FDS. While the connection between RGF distance and a change in the consensus sequence is intriguing, no significant genetic variation could be found in either dataset to explain the second peak. From the data studied here, the RGF distance may simply reflect the gradual accumulation of variation. Further longitudinal data with significant heterogeneity in variants is needed to study how local topological changes can be characterized using measures such as the RGF distance.

Fig. 2. RGF distance (y-axis) between consecutive graphs (x-axis) in FDS (solid) and IDS (dashed).

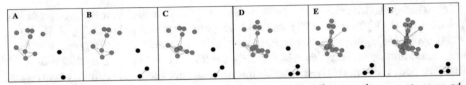

Fig. 3. Change in IDS graphs across time, including March (A) 19th, (B) 20th, (C) 21st, (D) 22nd, (E) 23rd, and (F) 24th. Nodes within the largest subnetwork are in blue.

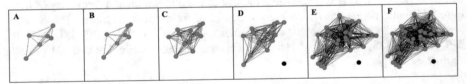

Fig. 4. Progression of the FDS connected component across February (A) 28th and (B) 29th, and March (C) 2nd, (D) 3rd, (E) 4th, and (F) 5th. Nodes of largest subnetwork are in blue.

3.4   Laplacian Network Partitioning Versus Phylogenetic Analysis

Initially, as new samples were added, the partitioning of IDS differed from graph to graph. The median normalized cut value between this time was 0.88 (IQR, 0.58–1.02). From March 30th through the remainder of the studied period, the partition in the graph remained the same. The partition consisted of two connected components with 164 and 5 samples, respectively, and the median normalized cut value decreased to 0.09 (IQR, 0.09–0.09). Samples in the latter component (EPI_ISL_447324, EPI_ISL_447322, EPI_ISL_447319, EPI_ISL_447271, and EPI_ISL_447265) shared 7 SNVs that were uncommon in the entire dataset. Those were G11083T (overall frequency, 0.06), C14805T (0.04), T17247C (0.04), C17676T (0.03), G26144T (0.04), G26660T (0.03), and C29627T (0.04). C17676T and G26660T were present only in this component and not in any sample in the larger component. None of the samples contained the 11 SNVs of interest. The samples that connected the small and large partitions were EPI_ISL_447271 and EPI_ISL_447447, respectively. EPI_ISL_447447 also had the highest betweenness centrality in the graph, as described in 3.2. The normalized cut value for the French dataset remained relatively consistent throughout the studied period (median = 1.18; IQR, 1–1.23). Additionally, until the last 3 days, the partitioning changed between graphs of different time periods. During these last three days, the cut, or the number of edges required to be removed to partition the graph, was 7.71% (638/8273), 6.72% (740/11011), and 6.6% (761/11532), respectively. The smaller partition consisted of 9 nodes (EPI_ISL_416746, EPI_ISL_414631, EPI_ISL_416494, EPI_ISL_418235, EPI_ISL_418426, EPI_ISL_418428, EPI_ISL_420047, EPI_ISL_420043, EPI_ISL_419178). 6 out of 9 samples shared mutations in the nucleocapsid phosphoprotein at positions G28881A, G28882A, G28883C. These mutations were not found in any sample in the larger connected component. In both IDS and FDS, the larger connected components had the C3037T, C14408T, A23403G mutations in over 95% and 100% of the samples, respectively. These mutations have been acknowledged as marker mutations of major clades and transmission clusters by GISAID and others [7, 30, 31], while the shared SNVS in the smaller IDS and FDS connected components have also been identified as markers of SARS-CoV-2 subtypes [32]. Partitioning of the last tracked graph, G_n, was compared to a phylogenetic tree representation of the data. As can be seen in Fig. 5, the importance of samples grouped together by spectral partitioning are less obvious when illustrated as a clade grouping on a phylogenetic tree and are even not necessarily within the same clade (Fig. 5B).

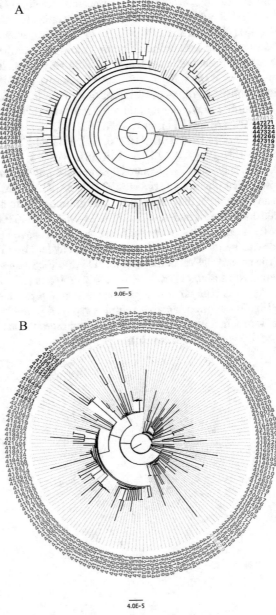

Fig. 5. Phylogenetic tree for (A) IDS and (B) FDS. Colors indicate samples grouped together by spectral partitioning (darkest and second darkest grey, respectively) and samples not part of the major subnetwork (light grey). Numberings refer to the GISAID sample accession numbers.

4   Conclusions

Here, we add a temporal dimension to a network representation to elucidate a connection between change in topological properties and viral molecular evolution within a population. Temporal dynamic networks, or time-varying graphs, can be represented as multiple networks acting as "snapshots" of the overall network changing in time.

In the two datasets studied by us, tracking of changes in the network reflected the evolution in the underlying viral genetic population. Small changes in global properties such as diameter and average path length were congruent with the low level of change in the overall nucleotide diversity of the population. Spectral partitioning of the graph was able to highlight communities of samples with shared SNVs not obvious from a phylogenetic construction of the data. The implication of graphlet-based analyses local topological analysis is less clear. Although preliminary results presented here found that the largest RGF distance between two temporally adjacent networks coincided with a shift in the consensus sequence of the population, this finding was not consistent in other temporally adjacent networks with a relatively large RGF distance.

Acknowledgements. This research was funded in part by National Science Foundation grant IIS-1817239.

Genome sequences analyzed is this work were submitted and collected be the following laboratories: IDS was submitted by the Stern Lab and collected by Microbiology laboratory, Assuta Ashdod University-Affiliated Hospital (EPI_ISL_447258 - 80); Microbiology Division, Barzilai University Medical Center (EPI_ISL_447281-310); Clinical Virology Laboratory, Soroka Medical Center and the Faculty of Health Sciences, Ben-Gurion University of the Negev (EPI_ISL_447311-30); Clinical Virology Unit, Hadassah Hebrew University Medical Center (EPI_ISL_447331-82, EPI_ISL_447407-16), Clinical Microbiology Laboratory, The Baruch Padeh Medical Center, Poriya (EPI_ISL_447383-406, EPI_ISL_447417-8); and Clinical Microbiology Laboratory, Sheba Medical Center (EPI_ISL_447419-69).

FDS was submitted by two laboratories, namely the National Reference Center for Viruses of Respiratory Infections, Institut Pasteur, Paris. Samples were collected by the Centre Hositalier Universitaire de Rouen Laboratoire de Virologie (EPI_ISL_414624, EPI_ISL_416494); Centre Hospitalier Régional Universitaire de Nantes Laboratoire de Virologie (EPI_ISL_414625); Centre Hospitalier Compiègne Laboratoire de Biologie (EPI_ISL_414627, EPI_ISL_414629-30, EPI_ISL_414634-8, EPI_ISL_415653-4, EPI_ISL_416495-7, EPI_ISL_418218, EPI_ISL_418220-1, EPI_ISL_418223-5, EPI_ISL_418227-8, EPI_ISL_418231, EPI_ISL_418236-9); Hôpital Robert Debré Laboratoire de Virologie (EPI_ISL_414631-2); Centre Hospitalier René Dubois Laboratoire de Microbiologie - Bât A (EPI_ISL_414633); Hôpital Instruction des Armées - BEGIN (EPI_ISL_415650); CH Jean de Navarre Laboratoire de Biologie (EPI_ISL_416493, EPI_ISL_420044, EPI_ISL_420053); Institut Médico legal - Hop R. Poincaré (EPI_ISL_416498); LABM GH nord Essonne (EPI_ISL_416498); Hopital franco britannique - Service des Urgences (EPI_ISL_416501); CHRU Pontchaillou - Laboratoire de Virologie (EPI_ISL_416502, EPI_ISL_416504-6, EPI_ISL_416508-13); CHU - Hôpital Cavale Blanche - Labo. de Virologie (EPI_ISL_418219); CHRU Bretonneau - Serv. Bacterio-Virol. (EPI_ISL_418222); EHPAD - Résidences les Cèdres (EPI_ISL_418226); Hopital franco britannique - Laboratoire (EPI_ISL_418229); Clinique AVERAY LA BROUSTE, Med. Polyvalente (EPI_ISL_418230); Service des Urgences (EPI_ISL_418232-3); Cabinet médical (EPI_ISL_418235); Sentinelles network (EPI_ISL_420038, EPI_ISL_420045, EPI_ISL_420055, EPI_ISL_421514), L'Air du Temps (EPI_ISL_420039-40); CH Compiègne Laboratoire de Biologie (EPI_ISL_420041, EPI_ISL_420049-50, EPI_ISL_420056-7, EPI_ISL_421500, EPI_ISL_421509-11); Service de Biologie clinique (EPI_ISL_420042, EPI_ISL_421513); CMIP (EPI_ISL_420043, EPI_ISL_420061); Résidence Villa Caroline (EPI_ISL_420046-7); Service de Biologie Médicale - BP 125 (EPI_ISL_420048, EPI_ISL_420058-60, EPI_ISL_420062, EPI_ISL_420064, EPI_ISL_421501, EPI_ISL_421504-6, EPI_ISL_421512), Résidence Eleusis (EPI_ISL_420051); Résidence les Marines (EPI_ISL_420052); Résidence de maintenon (EPI_ISL_420054); Labo BM - Site de Juvisy - Hopital Général (EPI_ISL_420063); Parc des Dames (EPI_ISL_421502-3); Le Château de Seine-Port (EPI_ISL_421507-8), and unknown (EPI_ISL_414626, EPI_ISL_415649, EPI_ISL_415651-2, EPI_ISL_415649).

The second submitting laboratory was CNR Virus des Infections Respiratoires - France SUD. Samples were collected by CNR Virus des Infections Respiratoires – France SUD (EPI_ISL_416745-6); Institut des Agents Infectieux (IAI) Hospices Civils de Lyon (EPI_ISL_416747-8, EPI_ISL_416750, EPI_ISL_416754, EPI_ISL_416756, EPI_ISL_416758); Centre Hospitalier de Valence (EPI_ISL_416749, EPI_ISL_418414-5, EPI_ISL_418417, EPI_ISL_419168); CHU Gabriel Montpied (EPI_ISL_416751-2); Centre Hospitalier de Bourg en Bresse (EPI_ISL_416757, EPI_ISL_417340, EPI_ISL_418426, EPI_ISL_419183, EPI_ISL_419185-6, EPI_ISL_420620); Institut des Agents Infectieux (IAI), Hospices Civils de Lyon (EPI_ISL_417333-4, EPI_ISL_417336-7, EPI_ISL_417339, EPI_ISL_418420-5, EPI_ISL_418429-31, EPI_ISL_419169-73, EPI_ISL_419177-82, EPI_ISL_419184, EPI_ISL_420604-11, EPI_ISL_420615-6, EPI_ISL_420618-9, EPI_ISL_420621-5); Centre Hospitalier de Macon (EPI_ISL_417338, EPI_ISL_418413, EPI_ISL_419174-6, EPI_ISL_419187-8, EPI_ISL_420612-4); Centre Hospitalier des Vals d'Ardeche (EPI_ISL_418412); GH Les Portes du Sud (EPI_ISL_418416); Centre Hospitalier Saint Joseph Saint Luc (EPI_ISL_418418-9, EPI_ISL_420617); Hopital Privé de l'Est Lyonnais (EPI_ISL_418418-9, EPI_ISL_420617); and Centre Hospitalier Lucien Hussel (EPI_ISL_418428).

References

1. World Health Organization. https://covid19.who.int
2. Shah, V.K., Firmal, P., Alam, A., Ganguly, D., Chattopadhyay, S.: Overview of immune response during SARS-CoV-2 infection: lessons from the past. Front. Immunol. **11**, 1949 (2020). https://doi.org/10.3389/fimmu.2020.01949
3. Peng, L., et al.: Prioritizing antiviral drugs against SARS-CoV-2 by integrating viral complete genome sequences and drug chemical structures. Sci. Rep. **11**, 6248 (2021). https://doi.org/10.1038/s41598-021-83737-5
4. Naqvi, A.A.T., et al.: Insights into SARS-CoV-2 genome, structure, evolution, pathogenesis and therapies: structural genomics approach. Biochim. Biophys. Acta Mol. Basis Dis. **1866**, 165878 (2020). https://doi.org/10.1016/j.bbadis.2020.165878
5. Miller, D., et al.: Full genome viral sequences inform patterns of SARS-CoV-2 spread into and within Israel. Nat. Commun. **11**, 5518 (2020). https://doi.org/10.1038/s41467-020-19248-0
6. Danesh, G., et al.: The COVID SMIT PSL group: early phylodynamics analysis of the COVID-19 epidemic in France. Epidemiology (2020). https://doi.org/10.1101/2020.06.03.20119925
7. Global Initiative on Sharing Avian Influenza Data (GISAID). https://www.gisaid.org
8. Wymant, C., et al.: STOP-HCV consortium, the maela pneumococcal collaboration, and the BEEHIVE collaboration: PHYLOSCANNER: inferring transmission from within- and between-host pathogen genetic diversity. Mol. Biol. Evol. **35**, 719–733 (2018). https://doi.org/10.1093/molbev/msx304

9. Sledzieski, S., Zhang, C., Mandoiu, I., Bansal, M.S.: TreeFix-TP: phylogenetic error-correction for infectious disease transmission network inference. In: Pacific Symposium on Biocomputing, vol. 26, pp. 119–130 (2021). https://doi.org/10.1142/9789811232701_0012

10. Didelot, X., Kendall, M., Xu, Y., White, P.J., McCarthy, N.: Genomic epidemiology analysis of infectious disease outbreaks using transphylo. Curr. Protoc. **1**, e60 (2021). https://doi.org/10.1002/cpz1.60

11. Volz, E.M., Koelle, K., Bedford, T.: Viral phylodynamics. PLoS Comput. Biol. **9**, e1002947 (2013). https://doi.org/10.1371/journal.pcbi.1002947

12. Nascimento, F.F., dos Reis, M., Yang, Z.: A biologist's guide to Bayesian phylogenetic analysis. Nat. Ecol. Evol. **1**, 1446–1454 (2017). https://doi.org/10.1038/s41559-017-0280-x

13. Zarrabi, N., Prosperi, M., Belleman, R.G., Colafigli, M., De Luca, A., Sloot, P.M.A.: Combining epidemiological and genetic networks signifies the importance of early treatment in HIV-1 transmission. PLoS ONE **7**, e46156 (2012). https://doi.org/10.1371/journal.pone.0046156

14. Kuchaiev, O., Stevanović, A., Hayes, W., Pržulj, N.: GraphCrunch 2: software tool for network modeling, alignment and clustering. BMC Bioinform. **12**, 24 (2011). https://doi.org/10.1186/1471-2105-12-24

15. Hayes, W., Sun, K., Przulj, N.: Graphlet-based measures are suitable for biological network comparison. Bioinformatics **29**, 483–491 (2013). https://doi.org/10.1093/bioinformatics/bts729

16. Skums, P., et al.: QUENTIN: reconstruction of disease transmissions from viral quasispecies genomic data. Bioinformatics **34**, 163–170 (2018). https://doi.org/10.1093/bioinformatics/btx402

17. Vecchio, F., Miraglia, F., Maria Rossini, P.: Connectome: graph theory application in functional brain network architecture. Clin. Neurophysiol. Pract. **2**, 206–213 (2017). https://doi.org/10.1016/j.cnp.2017.09.003

18. Cardoso, D.M., Delorme, C., Rama, P.: Laplacian eigenvectors and eigenvalues and almost equitable partitions. Eur. J. Comb. **28**, 665–673 (2007). https://doi.org/10.1016/j.ejc.2005.03.006

19. Kosakovsky Pond, S.L., Weaver, S., Leigh Brown, A.J., Wertheim, J.O.: HIV-TRACE (TRAnsmission Cluster Engine): a tool for large scale molecular epidemiology of HIV-1 and other rapidly evolving pathogens. Mol. Biol. Evol. **35**, 1812–1819 (2018). https://doi.org/10.1093/molbev/msy016

20. Campo, D.S., et al.: Accurate genetic detection of hepatitis C virus transmissions in outbreak settings. J. Infect. Dis. **213**, 957–965 (2016). https://doi.org/10.1093/infdis/jiv542

21. Poon, A.F.Y., et al.: Near real-time monitoring of HIV transmission hotspots from routine HIV genotyping: an implementation case study. Lancet HIV. **3**, e231-238 (2016). https://doi.org/10.1016/S2352-3018(16)00046-1

22. Bartenschlager, R., Lohmann, V.: Replication of hepatitis C virus. J Gen. Virol. **81**, 1631–1648 (2000). https://doi.org/10.1099/0022-1317-81-7-1631

23. Lorenzo-Redondo, R., Ozer, E.A., Achenbach, C.J., D'Aquila, R.T., Hultquist, J.F.: Molecular epidemiology in the HIV and SARS-CoV-2 pandemics. Curr. Opin. HIV AIDS **16**, 11–24 (2021). https://doi.org/10.1097/COH.0000000000000660

24. Katoh, K., Rozewicki, J., Yamada, K.D.: MAFFT online service: multiple sequence alignment, interactive sequence choice and visualization. Brief Bioinform. **20**, 1160–1166 (2019). https://doi.org/10.1093/bib/bbx108

25. Przulj, N., Corneil, D.G., Jurisica, I.: Modeling interactome: scale-free or geometric? Bioinformatics **20**, 3508–3515 (2004). https://doi.org/10.1093/bioinformatics/bth436

26. Nei, M., Li, W.H.: Mathematical model for studying genetic variation in terms of restriction endonucleases. Proc. Natl. Acad. Sci. **76**, 5269–5273 (1979). https://doi.org/10.1073/pnas.76.10.5269

27. Kozlov, A.M., Darriba, D., Flouri, T., Morel, B., Stamatakis, A.: RAxML-NG: a fast, scalable and user-friendly tool for maximum likelihood phylogenetic inference. Bioinformatics **35**, 4453–4455 (2019). https://doi.org/10.1093/bioinformatics/btz305

28. Ugurel, O.M., Ata, O., Turgut-Balik, D.: An updated analysis of variations in SARS-CoV-2 genome. Turk. J. Biol. **44**, 157–167 (2020). https://doi.org/10.3906/biy-2005-111

29. Zheng, Y.-X., et al.: Nsp2 has the potential to be a drug target revealed by global identification of SARS-CoV-2 Nsp2-interacting proteins. Acta Biochim. Biophys. Sin. **53**, 1134–1141 (2021). https://doi.org/10.1093/abbs/gmab088

30. Yang, X., Dong, N., Chan, E.W.-C., Chen, S.: Genetic cluster analysis of SARS-CoV-2 and the identification of those responsible for the major outbreaks in various countries. Emerg. Microbes Infect. **9**, 1287–1299 (2020). https://doi.org/10.1080/22221751.2020.1773745

31. Bai, Y., et al.: Comprehensive evolution and molecular characteristics of a large number of SARS-CoV-2 genomes reveal its epidemic trends. Int. J. Infect. Dis. **100**, 164–173 (2020). https://doi.org/10.1016/j.ijid.2020.08.066

32. Yang, H.-C., et al.: Analysis of genomic distributions of SARS-CoV-2 reveals a dominant strain type with strong allelic associations. Proc. Natl. Acad. Sci. U.S.A. **117**, 30679–30686 (2020). https://doi.org/10.1073/pnas.2007840117

An SVM Based Approach to Study the Racial Disparity in Triple-Negative Breast Cancer

Bikram Sahoo[(✉)][iD], Seth Sims[iD], and Alexander Zelikovsky[(✉)][iD]

Department of Computer Science, Georgia State University,
Atlanta, GA 30302, USA
biks.kestro@gmail.com, alex.zelikovsky@gmail.com

Abstract. Triple-negative breast cancer (TNBC) is one of the most heterogeneous molecular subtypes of breast cancer. TNBC is well-known for its poor survival rate, with limited treatment options compared to other breast cancer subtypes. The outcome of the disease is worse, especially in the case of African American women (AA) than European American women (EA), by showing the lowest survival rate in every stage of the disease. The current study considered a Support Vector Machine (SVM) based approach using gene expression data to find possible genes that can aid in understanding the molecular mechanism behind the racial disparity in triple-negative breast cancer. The implementation of linear kernel SVM in gene expression analysis already showed some promising results handling a large number of features/genes. This study found a unique set of genes that can accurately classify the gene expression data into two groups: EA and AA. To validate our results, we considered *KLK10*, which is highly expressed in AA women, and elevated *KLK10* is known for cancer progression and poor survival. Further improvement of the current method and analysis of our results is necessary to understand the racial disparity more accurately in TNBC.

Keywords: Triple negative breast cancer · Racial disparity · Machine learning · SVM · Feature engineering

1 Introduction

Breast cancer is the second-most common cancer in women and a significant cause of cancer-related death for women in the USA [16]. Three molecular markers classify breast cancer: estrogen receptor (ER), progesterone receptor (PR), and human epidermal growth factor receptor (Her2). In terms of bio-molecular classification, estrogen and progesterone are the hormones, and HER2 is the protein. Generally, breast cancer therapies focus on blocking or reducing these hormones and targeting HER2 protein [6,25]. Among all types of breast cancer, triple-negative breast cancer (TNBC) is a sub-type of breast cancer with the absence of estrogen, progesterone receptors, and HER2 proteins [29]. TNBC has

© The Author(s), under exclusive license to Springer Nature Switzerland AG 2022
M. S. Bansal et al. (Eds.): ICCABS 2021, LNBI 13254, pp. 163–175, 2022.
https://doi.org/10.1007/978-3-031-17531-2_13

highly heterogeneous cancer cells with different biological features. In TNBC, the lack of estrogen, progesterone, and HER2 protein forces doctors to use surgery, chemotherapy, and radiation. Because both hormonal and HER2 mediated therapies will not work in triple-negative breast cancer, the rate of triple-negative cancer diagnosis in the US only is 10–20% among all invasive breast cancer diagnoses [9,22,29]. According to epidemiological studies in breast cancer, it is well known that African American (AA) women have a higher risk of TNBC than European American (EA) women. The outcome of the study revealed that major contributing factors for this disparity in TNBC are biological and socioeconomic causes [5,6,26,32,34]. Current research requires robust computational models to process the clinical, epidemiological, and molecular data, which can retrieve the in-depth information to understand the factors causing the treatment and survival, which will help implement better treatment options in the clinics.

The reduced cost of next-generation sequencing technology has aided researchers in generating a massive amount of biological data to understand the molecular mechanism of the disease. RNA sequencing is one of the techniques of NGS technology used by researchers to study the gene expression, gene fusion, gene-gene interaction network, and pathway analysis between different types of biological conditions. Once a research group publishes the study, the data used in the study is freely available with all the biological and clinical information in various open-source databases [7,21,31]. Therefore, a massive amount of RNA-sequencing gene expression data is currently available on databases: NCBI-GEO, ArrayExpress, and ICGC [12,14,18]. To understand more about the diseases and resolve unsolved biological puzzles, there is an unmet need for state-of-art statistical and computational models to leverage this massive amount of RNA-seq data to generate biologically meaningful information.

This study considered a support vector machine (SVM) approach to process the open-source RNA sequencing gene expression data to find a significant set of genes that can classify AA and EA women and predict the survival outcome. In order to validate the final gene list generated from our model, we performed a cox proportional-hazards model and kaplan meier statistical test on clinical data. A linear kernel SVM can handle many features when building a classifier; in the case of RNA-seq data, features are genes. Studies considered the SVM-based approach to develop a model that can classify gene expression data and showed the advantage over traditional methods such as the unsupervised clustering approach [4,8,15,28,35].

Our analysis identified *KLK10* as a gene over-expressed in AA patients compared to EA patients. *KLK10* plays a significant role in cell progression. A survival analysis indicates that high *KLK10* expression is associated with poor prognosis in TNBC in our data set. High expression of *KLK10* is known more generally to be a signal of poor prognosis of TNBC cases [1,19,23,30,39]. We conclude from this that our methodology has successfully identified a significant biomarker.

2   Data and Methods

2.1   Data and Prepossessing

We downloaded the log2 median-centered RNA-seq gene expression data of breast cancer with clinical information [14, 18]. In order to get the triple-negative breast cancer samples, we considered IHC receptor status negative for ER, PR, and HER2 from the clinical data. In addition to that, we considered FISH status wherein HER2 status is equivocal for a breast cancer sample in the clinical data. This filtering based on receptor status helped us reduce our sample size to 145 by only having triple-negative breast cancer samples. Furthermore, we considered the race category information in the clinical data to collect White and Black/African American triple-negative breast cancer patients' RNA-seq gene expression data. Finally, we have 128 triple-negative breast cancer samples of white and black/African American races for this study.

2.2   Construction of SVM Based Model for Feature Selection

Features that had few unique counts across the data points were dropped. The expression count data was then standardised to Z-scores. A linear-kernel support vector machine (SVM) classifier was fitted to the transformed data. L1 regularization was used to produce a classifier for the groups with a sparse feature set as an initial feature selection method.

Spearman's rho was calculated pairwise between features and Ward's method linkage used to discover highly correlated groups of features. One feature from each cluster was selected at random to represent each cluster. We selected clusters using a threshold of 0.65 on the linkage-distance score. This value was selected by a grid search in 5 fold cross validation. The search encompassed values between 0.25 and 1.75 with a step-size of 0.1. A linear SVM with standard L2 loss was fit to each fold and the threshold was set to the largest value which maintained complete separation in all folds. Figure 4 plots the accuracy of each fold during the grid search.

The number of features were further reduced using recursive feature elimination with cross validation [15]. This method has been used successfully to find classification features for other cancers. We continued to use a linear kernel SVM as the core classifier for the feature selection.

A permutation test was performed on the final features. The EA/AA labels are randomly permuted and fed to the feature selection pipeline. The number of times the magnitude of a feature coefficient from the permuted data exceeded the features selected on the original data was counted. This was repeated with 5,000 permutations.

The influence of each feature was then assessed with Shapley Additive Explanation (SHAP) values [24]. Bootstrapping was used to assess the distribution of variation of SHAP values for each feature. Each iteration was performed on a randomly selected test-train split of the data. A linear SVM was fit with Platt scaling estimation of class probabilities [28]. The SHAP KernelExplainer method

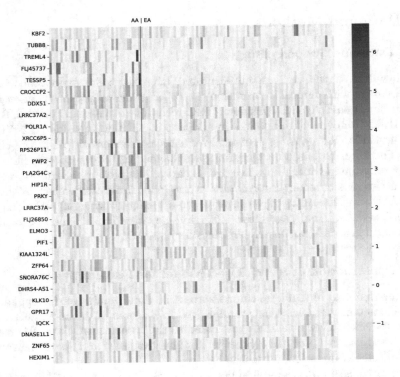

Fig. 1. The heatmap representation of the final genes selected by our SVM based model in AA and EA women.

Table 1. Cross validation of the final SVM

Fold	1st	2nd	3rd	4th	5th
Balanced Accuracy	1.0	0.94	0.96	1.0	1.0
ROC AUC	1.0	1.0	1.0	1.0	1.0
Weighted F1	1.0	0.96	0.96	1.0	1.0

was used to assess the influence of each item on the test set. The mean of the SHAP value magnitudes were collected, and plotted Fig. 2.

Finally each feature was evaluated using Student's T-test for independence of the mean with false discovery rate (FDR) control using the Benjamini-Hochberg procedure. Features with FDR rates > 0.05 were suppressed.

All feature selection was performed in a Jupyter notebook [20], using SciKit-learn [27], SciPy [36], Pandas [33], Seaborn [37], and Matplotlib [17].

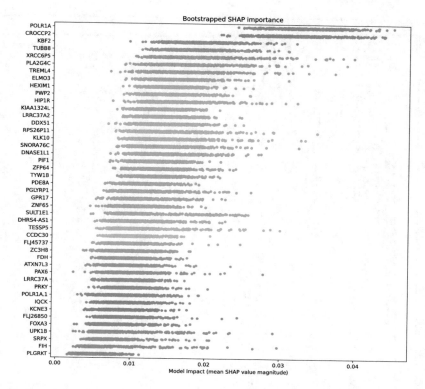

Fig. 2. Bootstrapped distribution of SHAP feature importance for each transcript. They are ordered by the median value of each bootstrap.

2.3 Estimation of Feature/Gene

Linear SVM optimize a the weights of a linear discriminant function.

$$D(\mathbf{x}) > 0 \Rightarrow \mathbf{x} \in \text{class } (+)$$
$$D(\mathbf{x}) < 0 \Rightarrow \mathbf{x} \in \text{class } (-)$$
$$D(\mathbf{x}) = 0, \text{ decision boundary}$$
$$\text{where}$$
$$D(\mathbf{x}) = \mathbf{w} \cdot \mathbf{x} + \mathbf{b}$$

Because of this simple form, the signs of the weight coefficients of the decision discriminant function can be directly interpreted as differences in expression. The signs represent the direction the positive class differs from the negative class. Survival analyses was performed using these expression differences.

2.4 Feature/Gene Validation Analysis

In order to validate the final gene list generated by our method, we considered a micro-array data set for TNBC with ER, PR and Her2 negative status

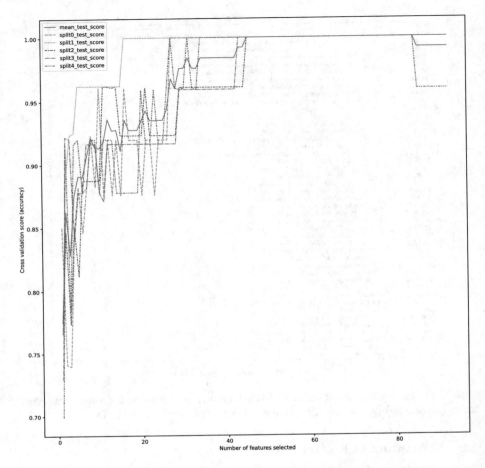

Fig. 3. Accuracy of SVM classification during recursive feature selection.

and survival data. Considering the cut-off value for each gene, we performed a Kaplan-Mier analysis and log-rank test. Further, we considered the genes with a p-value < 0.09 to discuss the TNBC racial disparity's biological role and build a gene signature.

3   Results and Discussion

3.1   Selection of Final Features

The data set contains 128 data points with transcriptomes of 87 European American women and 41 African American women. Each data point has 20,530 gene expression counts and initial inspection show that no single feature separate the classes. The initial selection using a L1 regularized classifier identified 95 features of interest. Performing Spearman's rho correlation combined with Ward's

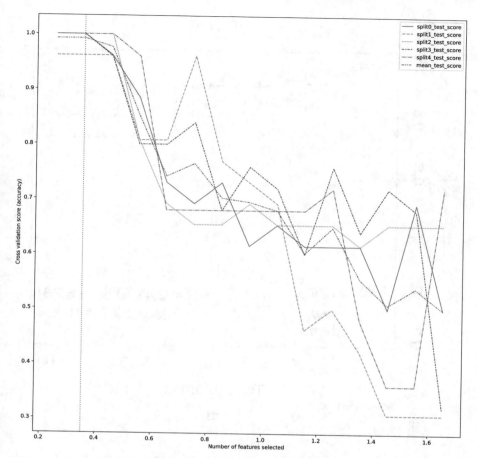

Fig. 4. Accuracy of SVM classification during grid search for collinearity threshold.

method at the chosen threshold yields 45 clusters. One feature from each cluster is selected to continue to recursive feature elimination with cross validation. The elimination procedure selected a final number of 24 features with a final mean accuracy of 98% Fig. 3.

Validation of Selected Features. Five fold cross validation of the final features yielded the scores in Table 1. The features allow for near perfect separation of the two classes. The bootstrap SHAP distribution presented in Fig. 2 indicate that the *CROCCP2* and *POLR1A* transcript has the largest influences on the model. The permutation test excluded all but four features at the 5% level. The four features were *POLR1A, CROCCP2, KBF2,* and *SULTIE1.*

While the near perfect separate of the classes was achieved, the large number of initial features may allow an uninformative set of features to partition the small number of data points. Also the classification of the points can only be a

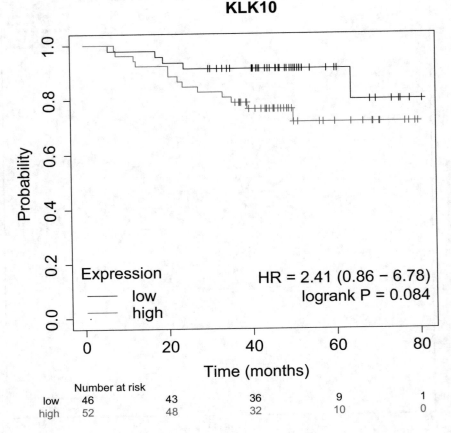

Fig. 5. High expression of KLK10 shows poor overall survival in the TNBC patients with p-value < 0.09 and Hazard ratio [HR] = 2.41 (0.86–6.78). In the above plot, red represents the TNBC samples with high KLK10 expression and black represents the TNBC samples with low KLK10 expression.

proxy for risk from TNBC. In order to further refine the analysis we use published survival information. We used the Kaplan-Mier analysis and log rank test to find associations between the discovered features and TNBC survival outcomes.

3.2   Expression Status of the Features/Gene

The differences in expression of the final gene list is in Fig. 1 and Table 2. *KLK10* is one of the gene which is highly expressed in AA samples in our analysis. In the subsequent survival analysis shows that the high expression of *KLK10* gene leads to poor survival in TNBC. Furthermore, the high expression of *KLK10* in tumors are already reported in various cancers including TNBC [1,19,23,30,39].

Table 2. Gene expression status. Each difference was investigated by Student's T-test for Independence of means with FDR correction by the Benjamini-Hochberg procedure.

	EA	AA	EA means	AA means	FDR	Reject
KBF2	H	L	746.90	525.83	<0.001	True
TUBB8	L	H	11.01	23.60	<0.001	True
TREML4	L	H	0.53	2.76	<0.001	True
FLJ45737	L	H	13.35	51.86	<0.001	True
TESSP5	L	H	1.99	10.31	<0.001	True
CROCCP2	L	H	234.93	520.63	<0.001	True
DDX51	L	H	411.13	721.80	<0.001	True
LRRC37A2	H	L	125.21	61.89	<0.001	True
POLR1A	H	L	407.11	161.32	<0.001	True
XRCC6P5	L	H	0.74	1.98	<0.001	True
RPS26P11	L	H	0.86	2.06	<0.001	True
PWP2	L	H	1062.20	1619.12	<0.001	True
PLA2G4C	L	H	132.70	279.76	<0.001	True
HIP1R	L	H	932.62	1620.51	<0.001	True
PRKY	L	H	2.26	5.93	0.001	True
LRRC37A	H	L	320.96	201.34	0.001	True
FLJ26850	L	H	0.95	3.64	0.001	True
ELMO3	L	H	381.44	679.34	0.001	True
PIF1	L	H	105.26	190.83	0.001	True
KIAA1324L	H	L	242.47	112.44	0.002	True
ZFP64	L	H	485.34	607.47	0.002	True
SNORA76C	L	H	0.13	0.45	0.003	True
DHRS4-AS1	H	L	649.67	388.31	0.003	True
KLK10	L	H	735.21	2746.76	0.005	True
GPR17	L	H	4.95	18.19	0.006	True
IQCK	H	L	408.64	284.22	0.025	True
DNASE1L1	L	H	472.65	661.69	0.028	True
ZNF65	H	L	1014.11	782.45	0.029	True
HEXIM1	L	H	782.62	987.46	0.029	True
FDH	H	L	2414.35	1998.87	0.063	False
KCNE3	H	L	154.85	112.82	0.063	False
CCDC30	H	L	55.81	36.68	0.065	False
PGLYRP1	L	H	0.17	0.56	0.112	False
TYW1B	L	H	93.65	126.90	0.125	False
PDE8A	H	L	561.40	455.48	0.148	False
PAX6	L	H	133.19	298.26	0.148	False
ATXN7L3	L	H	1471.25	1654.79	0.152	False
FIH	L	H	0.35	1.80	0.202	False
FOXA3	L	H	8.36	19.11	0.22	False
SULT1E1	L	H	9.34	39.47	0.22	False
UPK1B	L	H	11.14	39.22	0.297	False
PLGRKT	H	L	433.78	363.53	0.362	False
ZC3H8	L	H	182.43	208.53	0.362	False
POLR1A.1	L	H	1079.08	1224.81	0.362	False
SRPX	H	L	465.60	354.63	0.392	False

3.3 Association of High Expression of *KLK10* and Survival Outcome of TNBC

KLK10 is a non-classical member of the karllikrein-related peptidases (KLKs) family. Several KLK family genes played a major role in cancer progression and were reported in various studies. *KLK10* is also known as *NES1* (Normal Epithelial cell-Specific 1 gene) because of its expression in normal mammary epithelial cells. Studies have been published reporting the role of KLK10 as a potential biomarker in various cancer [3, 19]. In the case of ovarian cancer, *KLK10* has already been considered a diagnostic and prognostic biomarker with *CA125* [2, 10, 11, 13, 38]. Up-regulation of *KLK10* mRNA expression is also reported in pancreatic and colorectal cancer [1, 30]. However, in the case of breast cancer and testicular cancer, studies have shown the down-regulation of *KLK10* [19, 23, 39]. Specifically, the literature suggests that *KLK10* high-expression supports TNBC progression. The Kaplan-Meier curve in (Fig. 5) shows that *KLK10* high expression leads to poor survival in TNBC (Hazard ratio [HR] = 2.41, $p < 0.08$).

4 Conclusion

The critical challenge in treating triple-negative breast is its aggressive progression and limited therapeutic option. Recently many clinical and epidemiological studies revealed a significant racial disparity in TNBC between African American and European American women. And unfortunately, African American women have been shown poorer outcomes than European American women. However, these studies reveal that genetic and environmental factors are the primary fuel behind the aggressive progression of TNBC in African American women. Still, there is an unmet need to understand the comprehensive molecular mechanism behind the racial disparity in TNBC to develop an accurately targeted therapy option. The NGS technology and robust computational methods have played a significant role so far in unraveling the molecular heterogeneity of TNBC and sub-divided TNBC into several sub-groups at the molecular level. However, there is always ambiguity while choosing the computational method for a dataset. Even further requires sound statistical knowledge to narrow down the results from a thousand genes to a handful of genes. Our proposed methodology uses an SVM classifier on the RNA-Seq gene expression data of African American and European American women to generate a small gene list with 24 genes. The genes generated by the SVM classifier played a significant role in cancer progression and were reported in different studies. In order to validate our method, we considered an *KLK10* gene from our final gene list produced by the SVM classifier, which is up-regulated in African American women. The biological role of *KLK10* in various cancer progression and metastasis is already reported in several studies. Also, we show that the over-expression of *KLK10* in TNBC leads to significant poor survival, which is less explored for understanding the TNBC disparity in African-American women.

Availability of Data and Materials. Data sets and Jupyter notebook hosted at Github: https://github.com/xzy3/SVM-TNBC-racial-disparity.

Contributions. BS prepared the data set, validated and analysed ML results, and wrote the manuscript. SS performed data ML analysis, and wrote the manuscript. AZ supervised the project.

BS was partially supported by a GSU Molecular Basis of Disease Fellowship. AZ was partially supported from NSF Grant CCF-2212508, and NIH grant 1R21CA241044-01A1.

References

1. Alexopoulou, D.K., Papadopoulos, I.N., Scorilas, A.: Clinical significance of kallikrein-related peptidase (KLK10) mRNA expression in colorectal cancer. Clin. Biochem. **46**, 1453–1461 (2013). https://doi.org/10.1016/j.clinbiochem.2013.03.002

2. Batra, J., et al.: Kallikrein-related peptidase 10 (KLK10) expression and single nucleotide polymorphisms in ovarian cancer survival. Int. J. Gynecol. Cancer **20**, 529–536 (2010). https://doi.org/10.1111/igc.0b013e3181d9273e

3. Borgoño, C.A., Diamandis, E.P.: The emerging roles of human tissue kallikreins in cancer. Nat. Rev. Cancer **4**, 876–890 (2004). https://doi.org/10.1038/nrc1474

4. Brown, M.P.S., et al.: Knowledge-based analysis of microarray gene expression data by using support vector machines. Proc. Natl. Acad. Sci. **97**, 262–267 (2000). https://doi.org/10.1073/pnas.97.1.262

5. Chen, L., Li, C.I.: Racial disparities in breast cancer diagnosis and treatment by hormone receptor and HER2 status. Cancer Epidemiol. Biomarkers Prev. **24**(11), 1666–1672 (2015). https://doi.org/10.1158/1055-9965.EPI-15-0293

6. Cho, B., et al.: Evaluation of racial/ethnic differences in treatment and mortality among women with triple-negative breast cancer. JAMA Oncol. **7**(7), 1016–1023 (2021). https://doi.org/10.1001/jamaoncol.2021.1254

7. Conesa, A., et al.: A survey of best practices for RNA-seq data analysis. Genome Biol. **17** (2016). https://doi.org/10.1186/s13059-016-0881-8. https://www.ncbi.nlm.nih.gov/pmc/articles/PMC4728800/

8. Das, P., Roychowdhury, A., Das, S., Roychoudhury, S., Tripathy, S.: sigFeature: novel significant feature selection method for classification of gene expression data using support vector machine and t statistic. Front. Genet. **11** (2020). https://doi.org/10.3389/fgene.2020.00247

9. Dietze, E.C., Sistrunk, C., Miranda-Carboni, G., O'Regan, R., Seewaldt, V.L.: Triple-negative breast cancer in African-American women: disparities versus biology. Nat. Rev. Cancer **15**(4), 248–254 (2015). https://doi.org/10.1038/nrc3896

10. Dong, Y., Loessner, D., Irving-Rodgers, H., Obermair, A., Nicklin, J.L., Clements, J.A.: Metastasis of ovarian cancer is mediated by kallikrein related peptidases. Clin. Exp. Metastasis **31**(1), 135–147 (2013). https://doi.org/10.1007/s10585-013-9615-4

11. El Sherbini, M.A., Mansour, A.A., Sallam, M.M., Shaban, E.A., Shehab ElDin, Z.A., El-Shalakany, A.H.: KLK10 exon 3 unmethylated PCR product concentration: a new potential early diagnostic marker in ovarian cancer? - a pilot study. J. Ovarian Res. **11** (2018). https://doi.org/10.1186/s13048-018-0407-y

12. EMBL-EBI: ArrayExpress < EMBL-EBI (2019). https://www.ebi.ac.uk/arrayexpress/

13. Geng, X., et al.: Clinical relevance of kallikrein-related peptidase 9, 10, 11, and 15 mRNA expression in advanced high-grade serous ovarian cancer. PLOS ONE **12**, e0186847 (2017). https://doi.org/10.1371/journal.pone.0186847

14. GEO: Home - GEO - NCBI (2019). https://www.ncbi.nlm.nih.gov/geo/

15. Guyon, I., Weston, J., Barnhill, S., Vapnik, V.: Gene selection for cancer classification using support vector machines. Mach. Learn. **46**(1), 389–422 (2002)

16. Hendrick, R.E., Helvie, M.A., Monticciolo, D.L.: Breast cancer mortality rates have stopped declining in U.S. women younger than 40 years. Radiology **299**, 143–149 (2021). https://doi.org/10.1148/radiol.2021203476

17. Hunter, J.D.: Matplotlib: A 2D graphics environment. Comput. Sci. Eng. **9**(3), 90–95 (2007). https://doi.org/10.1109/MCSE.2007.55

18. North Carolina Institute: The cancer genome atlas program (2019). https://www.cancer.gov/about-nci/organization/ccg/research/structural-genomics/tcga

19. Kioulafa, M., Kaklamanis, L., Stathopoulos, E., Mavroudis, D., Georgoulias, V., Lianidou, E.: Kallikrein 10 (KLK10) methylation as a novel prognostic biomarker in early breast cancer. Ann. Oncol. **20**, 1020–1025 (2009). https://doi.org/10.1093/annonc/mdn733

20. Kluyver, T., et al.: Jupyter notebooks - a publishing format for reproducible computational workflows. In: Loizides, F., Schmidt, B. (eds.) Positioning and Power in Academic Publishing: Players, Agents and Agendas, pp. 87–90. IOS Press (2016)

21. Lachmann, A., et al.: Massive mining of publicly available RNA-seq data from human and mouse. Nat. Commun. **9** (2018). https://doi.org/10.1038/s41467-018-03751-6

22. Lehrberg, A., et al.: Outcome of African-American compared to White-American patients with early-stage breast cancer, stratified by phenotype. Breast J. **27**(7), 573–580 (2021). https://doi.org/10.1111/tbj.14225

23. Lin, C.L., et al.: Transcriptional suppression of miR-7 by MTA2 induces Sp1-mediated KLK10 expression and metastasis of cervical cancer. Mol. Ther. - Nucleic Acids **20**, 699–710 (2020). https://doi.org/10.1016/j.omtn.2020.04.009

24. Lundberg, S.M., Lee, S.I.: A unified approach to interpreting model predictions. In: Guyon, I., Luxburg, U.V., et al. (eds.) Advances in Neural Information Processing Systems, vol. 30, pp. 4765–4774. Curran Associates, Inc. (2017). http://papers.nips.cc/paper/7062-a-unified-approach-to-interpreting-model-predictions.pdf

25. Moss, J.L., Tatalovich, Z., Zhu, L., Morgan, C., Cronin, K.A.: Triple-negative breast cancer incidence in the United States: ecological correlations with area-level sociodemographics, healthcare, and health behaviors. Breast Cancer **28**(1), 82–91 (2020). https://doi.org/10.1007/s12282-020-01132-w

26. Newman, L.A., Kaljee, L.M.: Health disparities and triple-negative breast cancer in African American women: a review. JAMA Surg. **152**(5), 485–493 (2017). https://doi.org/10.1001/jamasurg.2017.0005

27. Pedregosa, F., et al.: Scikit-learn: machine learning in Python. J. Mach. Learn. Res. **12**(Oct), 2825–2830 (2011)

28. Platt, J., et al.: Probabilistic outputs for support vector machines and comparisons to regularized likelihood methods. Adv. Large Margin Classifiers **10**(3), 61–74 (1999)

29. Prakash, O., Hossain, F., Danos, D., Lassak, A., Scribner, R., Miele, L.: Racial disparities in triple negative breast cancer: a review of the role of biologic and non-biologic factors. Front. Public Health **8** (2020). https://doi.org/10.3389/fpubh.2020.576964

30. Rückert, F., et al.: Co-expression of KLK6 and KLK10 as prognostic factors for survival in pancreatic ductal adenocarcinoma. Br. J. Cancer **99**, 1484–1492 (2008). https://doi.org/10.1038/sj.bjc.6604717

31. Rondel, F.M., et al.: Pipeline for analyzing activity of metabolic pathways in planktonic communities using metatranscriptomic data. J. Comput. Biol. **28**(8), 842–855 (2021). https://doi.org/10.1089/cmb.2021.0053

32. Siddharth, S., Sharma, D.: Racial disparity and triple-negative breast cancer in African-American women: a multifaceted affair between obesity, biology, and socioeconomic determinants. Cancers **10**(12) (2018). https://doi.org/10.3390/cancers10120514

33. jbrockmendel et al.: pandas-dev/pandas: Pandas (2020). https://doi.org/10.5281/zenodo.3509134

34. Sturtz, L.A., Melley, J., Mamula, K., Shriver, C.D., Ellsworth, R.E.: Outcome disparities in African American women with triple negative breast cancer: a comparison of epidemiological and molecular factors between African American and Caucasian women with triple negative breast cancer. BMC Cancer **14** (2014). https://doi.org/10.1186/1471-2407-14-62

35. Vanitha, C.D.A., Devaraj, D., Venkatesulu, M.: Gene expression data classification using support vector machine and mutual information-based gene selection. Procedia Comput. Sci. **47**, 13–21 (2015). https://doi.org/10.1016/j.procs.2015.03.178

36. Virtanen, P., et al.: SciPy 1.0: fundamental algorithms for scientific computing in Python. Nat. Methods **17**, 261–272 (2020). https://doi.org/10.1038/s41592-019-0686-2

37. Waskom, M.L.: Seaborn: statistical data visualization. J. Open Source Softw. **6**(60), 3021 (2021). https://doi.org/10.21105/joss.03021

38. White, N.M.A., et al.: Three dysregulated miRNAs control kallikrein 10 expression and cell proliferation in ovarian cancer. Br. J. Cancer **102**, 1244–1253 (2010). https://doi.org/10.1038/sj.bjc.6605634

39. Yousef, G.M., et al.: Human tissue kallikreins: from gene structure to function and clinical applications. Adv. Clin. Chem. 11–79 (2005). https://doi.org/10.1016/s0065-2423(04)39002-5

Author Index

Printed in the United States
by Baker & Taylor Publisher Services